VALERY A. KHOLODNYI received his PhD in Applied Mathematics from the Moscow Institute of Electronics and Mathematics in 1990. He has held university positions in various departments, such as the Department of Microwave and Quantum Electronics, the Department of Mathematical Modeling of Physical Systems, and the Department of Physics, in both Russia and the United States. He has authored or co-authored over 60 research papers in finance, mathematics, theoretical physics and engineering, and has published in journals such as the *Journal of Mathematical Physics* and the *Journal of Integral Equations and Applications*. He was an Invited Speaker at the Second World Congress of Nonlinear Analysts and at numerous international and national conferences, as well as at research seminars in university departments and industry. Currently, he is the Vice-President of Research and Development for Integrated Energy Services L.C., an independent research institute for financial capital markets.

JOHN F. PRICE received his PhD in Mathematics from the Australian National University in 1970 and has since held university positions in many countries, including Australia, Cambodia, Canada, England, Italy, Switzerland and the United States. He has published over 60 papers in mathematics, physics and finance in journals such as *Advances in Mathematics*, the *Journal of Functional Analysis* and *Notices of the American Mathematical Society*. He has also written articles for financial practitioners in *Risk*, *Export Today* and *Derivatives Strategy*, and has designed and implemented risk management software for major corporations.

Foreign Exchange Option Symmetry

VALERY A. KHOLODNYI
Integrated Energy Services L.C.
Iowa, USA

JOHN F. PRICE
Maharishi University of Management
Iowa, USA

World Scientific
Singapore • New Jersey • London • Hong Kong

Published by

World Scientific Publishing Co. Pte. Ltd.

P O Box 128, Farrer Road, Singapore 912805

USA office: Suite 1B, 1060 Main Street, River Edge, NJ 07661

UK office: 57 Shelton Street, Covent Garden, London WC2H 9HE

Library of Congress Cataloging-in-Publication Data
Kholodnyi, Valery A., 1964–
    Foreign exchange option symmetry / by Valery A. Kholodnyi and John
F. Price.
        p.   cm.
    Includes bibliographical references and index.
    ISBN 9810233620 (alk. paper)
    1. Foreign exchange market.   2. Options (Finance)   3. Derivative
securities.   4. Financial futures.   5. Symmetry groups.   I. Price,
John F.   II. Title.
HG3851.K3956   1998
332.4'5--dc21   97-43965
                                                            CIP

British Library Cataloguing-in-Publication Data
A catalogue record for this book is available from the British Library.

Printed in Singapore by Uto-Print

Preface

In 1900 Louis Bachelier, motivated by the problem of modeling the random character of the dynamics of prices of underlying securities in financial markets with the aim of analyzing financial derivatives, invented the seeds of stochastic calculus. Roughly speaking, financial derivatives are contracts whose specifications are determined by, that is, are derived from the prices of the underlying securities, such as stock or foreign currency. Although Bachelier's work lay relatively dormant for nearly half a century, his pioneering application of stochastic calculus, through the work of giants in the field of finance, culminated in the celebrated option pricing model of Fischer Black and Myron Scholes in 1973. Options are one of the most widely traded types of financial derivatives. Black and Scholes' groundbreaking work met with overwhelming success among both practitioners and academic researchers and presented a landmark understanding of this field. The Black and Scholes model was so successful, even too successful, in the sense that its very success led the study of actual financial phenomena, which was the original goal, to be virtually identified and later even replaced by the study of mathematical aspects of stochastic calculus, which was just the tool. The means were mistaken for the ends. The subject of financial derivatives literally became a subset of stochastic calculus.

In this book, we make an attempt to come back to the original goal, namely, to the study of actual financial phenomena, in particular, the financial phenomenon underlying the valuation of financial derivatives. It is our view that the study of financial phenomena is on the brink of a revolution similar to that of quantum physics in the 1920's, which was based on the study of actual physical phenomena that led to the thorough review and renewal of the very foundations of physics. It is this return to foundations that led, in turn, to the invention and elaboration of one of the most powerful and rich areas of mathematics and mathematical physics in this century.

Moreover, history shows that virtually all the major revolutions in physics that revealed the deepest insights into Nature were made through recognizing that a range of seemingly different phenomena actually represented a common underlying phenomenon. Such a unification indicates the presence of an inherent symmetry in the underlying phenomenon. A standard way to formalize such a symmetry mathematically is in terms of group theory.

In this book, we introduce a fundamental symmetry in a foreign exchange market that associates financially equivalent options on opposite sides of the market. We use this to establish symmetry relationships for the values of a wide range of foreign exchange options including European options with general payoffs; Bermudan and American options with general time-dependent payoffs; and barrier options with time-dependent, discretely and continuously activated barriers on general European, Bermudan, and American options. In the particular cases of European and American call and put options, these symmetry relationships take the form of a conjecture due to J. Orlin Grabbe in 1983 that such a call or put option with strike X on one side of a foreign exchange market is equivalent to a portfolio of X put or call options with strike $1/X$ on the opposite side of the market. Similar symmetry relationships are also valid for Bermudan call and put options.

The symmetry relationships hold in a general foreign exchange market environment. In particular, they require no assumptions on the nature of a probability distribution for exchange rates. In fact, we do not even assume the existence of such a distribution.

The symmetry at the level of foreign exchange options is based on the symmetry at the level of payoffs that associates with any given payoff, as a function of the corresponding exchange rate, on one side of the foreign exchange market a financially equivalent payoff on the opposite side of the market. These equivalent payoffs represent the same amount of wealth contingent upon the state of the market as expressed by the values of the exchange rates. It is this wealth that has an objective financial status and hence, using the terminology of modern theoretical physics, is a financial observable. In this regard, this wealth could be thought of as an abstract payoff, which is an element of the real vector space, or more precisely, the vector lattice of all abstract payoffs, while the actual payoffs, which could be received on either side of the foreign exchange market, are simply different viewpoints on the same amount of wealth.

We formalize the act of observing an abstract payoff, via its actual payoff on either the domestic or foreign side of the market, as a particular choice of a basis, the domestic or foreign basis, for the space of all abstract payoffs. In the chosen domestic or foreign basis, the coordinates of the abstract payoff, indexed by the corresponding exchange rate, are the values of its domestic or foreign actual payoff for this exchange rate. The concept of an abstract payoff, giving rise to a concept of an abstract option, allows for a coordinate-free description of financial phenomena in a foreign exchange market, that is, it allows for a description that is independent of the domestic or foreign basis. In this formalism the act of changing sides in a foreign exchange market is described as a change of these bases. The role of the change-of-basis operator is played by the one-dimensional Kelvin transform. It is in terms of this operator that we formulate our symmetry in a foreign exchange market.

Conceptually, our book is based on an article by one of the authors, Valery Kholodnyi, in 1995 (see [13]), which, to the best of our knowledge, introduced for the first time to finance such fundamental notions from modern theoretical physics as observables, invariant or coordinate-free descriptions of phenomena, symmetries and related group-theoretical methods.

In this regard, the book is in the light of development of the methods of the science of Theoretical Finance formulated by Kholodnyi and announced by him at the Special Session on Derivatives and Financial Mathematics of the 909th Meeting of the American Mathematical Society held in Iowa City, Iowa and at the Second World Congress of Nonlinear Analysts held in Athens, Greece both in 1996. Briefly speaking, Theoretical Finance is related to Finance in a similar manner as Theoretical Physics is related to Physics.

The framework of a market environment, which allows us to handle the symmetry relationships at the level of foreign exchange options without assumptions on the nature of a probability distribution for exchange rates, and, in fact, without the assumption of the existence of such a distribution, was introduced by Kholodnyi in 1995 (see [18]).

We freely make use of concepts, terminology, and results from these two articles in this book.

The book itself is based on our preprint [26], and a summary of the results appears in the proceedings of the Conference on Computational Intelligence for Financial Engineering held in New York in 1997 (see [28]).

In this book, for the sake of clarity of the exposition we restrict the presentation of our symmetry to a foreign exchange market with only two currencies. The symmetry in the case of a foreign exchange market with multiple currencies is presented in our preprint [27] and a summary of the results appears in our article [29].

The practical applications of the symmetry relationships presented in this book are significant and far reaching. We mention a few of the main applications here.

- The symmetry relationships provide practical criteria for detecting a completely new type of true arbitrage in foreign exchange markets;

- They provide a means of selecting, for a given foreign exchange portfolio, which may include options, a financially equivalent portfolio on the opposite side of a foreign exchange market, when for some reason it is not convenient or not possible to achieve the original portfolio;

- They provide a screen for testing consistency of option pricing models;

- The symmetry relationships also reduce significantly the cost and time of developing algorithms and software for valuing and analyzing portfolios of foreign exchange options and their Greek letters, for example, delta and gamma;

- The symmetry relationships are not limited to foreign exchange markets and, in fact, remain valid for any financial markets with exchange of arbitrary underlying securities.

Our book is written for a wide audience covering the whole spectrum of readers from financial practitioners to researchers in the areas of finance, mathematics, or physics:

- Relative value traders, who would like to take advantage of a completely new type of true arbitrage that is easy to execute;

- Independent private traders with minimal experience and limited technical and financial resources;

- Financial advisors and practitioners seeking a handbook, who need to apply the most up-to-date methods in finance in their daily professional activities in order to maintain their competitive edge, either personally or for their institutions;

- Financial professionals with limited background in quantitative finance, who seek to stay up to date with current progress in modern finance, or with minimal background in mathematics, who do not wish to be cumbered with mathematical proofs but seek deeper understanding of current trends in mathematical finance;

- Ph.D.s and other researchers in finance and mathematical finance;

- Mathematicians and physicists with or without knowledge of finance, who are interested in entering the field of modern mathematical finance;

- Students and teachers of finance, mathematics, or physics who would enjoy taking or teaching a course in the field of modern finance, especially in view of the current trend towards interdisciplinary programs in quantitative finance and risk management;

- Anyone with minimal background in mathematics or finance or both, who is interested in recent developments in modern finance.

To achieve these ends we first present our results in Part I, without formal mathematical proofs, using financial justification whenever possible. Second, we present in Part II the formal mathematical proofs of the results in Part I. Note that even in Part II we have chosen to leave out some of the more subtle mathematical details, such as the exact domains and ranges of the evolution operators defining the market environments, partly because we do not wish to overload the book with too many fine details only of interest to a pure mathematician, but mainly because these technical details may obscure the structure of the symmetry, which is purely algebraic in nature.

Those who are not interested in the mathematical proofs, such as financial advisors and practitioners, may choose to omit Part II. Even a minimal background of basic courses in calculus and linear algebra should take the reader to the cutting edge of understanding the financial principles underlying the symmetry relationships presented in this book. Furthermore, as much as possible we give complete lists of symmetry relationships from both sides of the foreign exchange market, so that the formulas can directly and easily be implemented as in a handbook, without any theoretical knowledge or extra reading.

Also, for the reader who wants to take the next step beyond simple application of formulas, we present whenever possible the symmetry relationships for European call and put options in the Black and Scholes market environment, where the values of these call and put options are given by the well-known Black and Scholes formulas. In this case we give direct proofs independent of the new theory.

Ph.D.s and other researchers in finance and financial mathematics may choose to read Part II or not, according to their taste.

Those mathematicians, physicists, and other readers with little knowledge of finance are referred to the review article by one of the authors, John Price, in 1996 (see [32]) for an introduction to basic financial terminology. With the help of this review article, those with sufficient mathematical background may choose to read the entire book and the rest just Part I.

Our book is organized as follows. We begin with an Introduction, which is Chapter 1.

Part I opens with Chapter 2, in which we introduce the notation, present the framework of a market environment used throughout the book, and give examples of two major market environments: the binomial market environment and the Black and Scholes market environment in the foreign exchange setting.

In Chapter 3, we present the symmetry between payoffs via the Kelvin transform at the heart of the theory and use it to derive basic symmetry relationships for European, Bermudan, and American options with general, and, in the latter two cases, time-dependent payoffs. Throughout, we show that in the particular cases of European and American call and put options, these symmetry relationships take the form of Grabbe's conjecture for such call and put options on opposite sides of the foreign exchange market, namely relationships (1.1) and (1.2) of Chapter 1. We also show that these relationships (1.1) and (1.2) hold as well for Bermudan call and put options.

In Chapter 4 we turn to exotic options, presenting symmetry relationships for one of the most widely traded types of exotic options, namely, barrier options. Here we consider barrier options with time-dependent, discretely and continuously activated barriers on general European, Bermudan, and American options. We then present symmetry relationships for Greek letters of foreign exchange options, in particular, for delta and gamma, and symmetry relationships for European, Bermudan, and American call and put options in the particular case of a market environment which we call the exchange-rate homogeneous market environment.

In Chapter 5 we describe how the symmetry can guide the choice of new payoffs that not only smooth the corners of the payoffs for the standard European, Bermudan, and American call and put options to improve their hedging characteristics, but also preserve the symmetry relationships satisfied by the call and put options themselves.

Part I ends with the discursive Chapter 6, in which we give an overview of the main practical applications of our symmetry.

In Part II Chapters 7 through 9 we present formal mathematical proofs of the results in Part I.

Finally, we would like to acknowledge the great help we received at all the stages of working on the book.

We thank, first of all, Integrated Energy Services, L.C. for financially supporting the realization of this book and for the use of its computer facilities, but most importantly for creating a unique environment that brought together practitioners and researchers. In our opinion, only in such a challenging and stimulating mixture of viewpoints can fresh and fruitful ideas, approaches, and solutions arise. The work that is the subject of this book represents the kernel of one of the many intellectual products for risk management offered by this company.

We thank Steve Rideout for his constant attention to the book and for the many practical insights into the nature of financial markets.

We thank Ralph McKay for valuable discussions. After we had introduced and proved the symmetry relationships for general European, Bermudan, and American options in the general market environment, he brought to our attention an alternative financial indication of the validity of our symmetry relationships in the particular case of European and American call and put foreign exchange options, which we later independently found to have been the subject of Grabbe's conjecture in 1983.

We thank George Paslaski and Geoffrey Golner for carefully reading the manuscript, and Sunil Rawal for his help in typesetting the manuscript.

We thank Anne Dow for independently verifying some of the formulas included in this book.

We thank The Mathworks for supplying us with copies of Matlab which we used to produce one of our charts.

One of us (Valery Kholodnyi) thanks his parents Antonina and Alexander for sharing their fascination with the world around us, what it is made from and how it works, but more importantly for their sleepless nights, care and understanding.

Saving the best for last, we thank our wives Larisa Kholodnaya and Sandy Price for their love, patience and actual help in preparing the book.

Contents

xiii

Contents

List of Figures

Foreign Exchange
Option Symmetry

Chapter 1

Introduction

It is well known that, due to a general no-arbitrage argument, there is a pairing for European call and put options referred to as put-call parity. In the foreign exchange setting this pairing has the form:

$$C_d(t, T, S, X) - P_d(t, T, S, X) = B_f(t, T)\,S - B_d(t, T)\,X,$$
$$C_f(t, T, S', X') - P_f(t, T, S', X') = B_d(t, T)\,S' - B_f(t, T)\,X',$$

where $C_d(t, T, S, X)$ and $P_d(t, T, S, X)$ are the values at inception time t of European call and put options in the domestic currency d on a foreign currency f with strike price X, expiration time T, and exchange rate S between currency d and currency f at time t, and where $B_d(t, T)$ and $B_f(t, T)$ are the values at inception time t of pure discount bonds with face values of one unit of the currencies d and f respectively at their maturity time T. Similarly, $C_f(t, T, S', X')$ and $P_f(t, T, S', X')$ are the values at inception time t of European call and put options in a foreign currency f on the domestic currency d with strike price X', expiration time T, and exchange rate S' between currency f and currency d at time t. (See, for example, Gibson [8].)

This type of pairing, limited to European call and put options, links call and put options on the same side of a foreign exchange market. There is, however, another possibly less known type of put-call option pairing that links call and put options on opposite sides of a foreign exchange market. This pairing holds not only for European, but also for American, call and put options. To the best of our knowledge, this latter type of pairing was first conjectured by Grabbe [9] who gave a certain financial indication of its plausibility for both European and American call and put options. Later Ritchken [33] also discussed its validity

1

for European call and put options. This type of pairing has the form:

$$(1.1) \qquad \begin{aligned} C_d(t,T,S,X) &= SX\,P_f(t,T,1/S,1/X), \\ P_d(t,T,S,X) &= SX\,C_f(t,T,1/S,1/X), \end{aligned}$$

$$(1.2) \qquad \begin{aligned} C_f(t,T,S',X') &= S'X'\,P_d(t,T,1/S',1/X'), \\ P_f(t,T,S',X') &= S'X'\,C_d(t,T,1/S',1/X'). \end{aligned}$$

where, in these relationships call and put options are respectively either European or American, and where by the no-arbitrage argument the exchange rates S and S' are assumed to be inverses to each other. In words, such a call or put option with strike X on the domestic side of a foreign exchange market is equivalent to a portfolio of X put or call options, respectively, with strike $1/X$ on the foreign side of the market. Similarly, a call or put option with strike X' on the foreign side of a foreign exchange market is equivalent to a portfolio of X' put or call options, respectively, with strike $1/X'$ on the domestic side of the market. We say that two options or two portfolios of options on the opposite sides of the foreign exchange market are equivalent, or more precisely, financially equivalent if converted to either side of the market they have the same values for all exchange rates.

In this book, we introduce and study a more fundamental symmetry in a general foreign exchange market environment with two currencies, domestic and foreign, which yields the preceding relationships (1.1) and (1.2) as particular cases.

Conceptually, our book is based on an article by Kholodnyi in 1995 (see [13]), which, to the best of our knowledge, introduced for the first time to finance such fundamental notions from modern theoretical physics as observables, invariant or coordinate-free descriptions of phenomena, symmetries and related group-theoretical methods.

The framework of a market environment, which allows us to handle the symmetry relationships at the level of foreign exchange options without assumptions on the nature of a probability distribution for exchange rates, and, in fact, without the assumption of the existence of such a distribution, was introduced by Kholodnyi in 1995 (see [18]).

We freely make use of concepts, terminology, and results from these two articles in this book.

Our symmetry in a foreign exchange market at the level of payoffs associates with any given payoff contingent upon the state of the market as expressed by the value of the corresponding exchange rate, a financially equivalent payoff on the opposite side of the market. We will now show that this association is accomplished by the Kelvin transform.

The (one-dimensional) *Kelvin transform* $\mathbf{K}_c : \Pi \to \Pi$, with Π being the vector space of all real-valued functions on the set of positive real numbers, is defined by

$$(\mathbf{K}_c\, g)(x) = (x/\sqrt{c})g(c/x), \quad x > 0,$$

where $c > 0$. When $c = 1$, we write $\mathbf{K}_c$ simply as $\mathbf{K}$.

Assume that we can receive an amount of wealth in either the domestic side or the foreign side of the foreign exchange market contingent upon the current state of the foreign exchange market as expressed by the values of the exchange rates. This wealth, which could be positive or negative, is independent of the particular currency in which it is measured. If the wealth is to be received in the domestic side of the market, that is, in currency d, contingent upon the exchange rate S between currency d and currency f, then the wealth is a payoff $h_d(S)$ in currency d as a function of the exchange rate S. Similarly, if the wealth is to be received in the foreign side of the market, that is, in currency f, contingent upon the exchange rate S' between currency f and currency d, then the wealth is a payoff $h_f(S')$ in currency f as a function of the exchange rate S'. Because h_d and h_f are real-valued functions on the set of positive real numbers, they are members of the function space Π.

Since the payoffs h_d and h_f express the same amount of wealth, they cannot be independent. They are, in fact, related via the Kelvin transform:

(1.3) $$h_d = \mathbf{K}\, h_f, \quad h_f = \mathbf{K}\, h_d.$$

These are the sought after symmetry relationships in the foreign exchange market at the level of payoffs. The following simple financial argument shows why these symmetry relationships hold.

Assume that we receive in the foreign side of the foreign exchange market a payoff $h_f(S')$ in currency f. By converting this payoff to currency d, we receive the payoff $S\, h_f(S')$. By the no-arbitrage argument, $S' = 1/S$ and so we can express the payoff $S\, h_f(S')$ solely in terms of S as $S\, h_f(1/S)$. Since the payoffs h_d and h_f express the same amount of wealth, the payoffs $h_d(S)$ and $S\, h_f(1/S)$ also express the same amounts of wealth, but now both are received on the same side of the foreign exchange market, the domestic side, and both are contingent upon the same exchange rate S. Therefore, because of the no-arbitrage argument, we obtain $h_d(S) = S\, h_f(1/S)$ for all $S > 0$. Since, by the definition of the Kelvin transform, $S\, h_f(1/S) = (\mathbf{K}h_f)(S)$ for all $S > 0$, we obtain the first equality of (1.3). By a similar argument, assuming that we receive in the domestic side of the foreign exchange market a payoff $h_d(S)$ in currency d, we obtain the second equality of (1.3).

In the particular case of the payoffs of call and put options the symmetry relationships in (1.3) take the form

$$\mathbf{K}\,(\cdot - X)^+ = X(1/X - \cdot)^+,$$
$$\mathbf{K}\,(X - \cdot)^+ = X(\cdot - 1/X)^+,$$

and

$$\mathbf{K}\,(\cdot - X')^+ = X'(1/X' - \cdot)^+,$$
$$\mathbf{K}\,(X' - \cdot)^+ = X'(\cdot - 1/X')^+,$$

where

$$(S - X)^+ \quad \text{and} \quad (X - S)^+$$

are the call and put payoffs in currency d with strike price X, and

$$(S' - X')^+ \quad \text{and} \quad (X' - S')^+$$

are the call and put payoffs in currency f with strike price X', and

$$x^+ = \begin{cases} x & \text{if } x \geq 0, \\ 0 & \text{if } x < 0. \end{cases}$$

These ideas can be formulated in the following natural abstract framework suggested by modern theoretical physics. Even though the wealth itself can only be directly observed through payoffs h_d and h_f, it is the wealth that has an objective financial status and hence, using the terminology of modern theoretical physics, is a *financial observable*. In this regard, this wealth could be thought of as an abstract payoff h in the space of all abstract payoffs $\mathbf{\Pi}$ in contrast to the payoffs h_d and h_f, that are actual payoffs. It is easy to see that the space of all abstract payoffs $\mathbf{\Pi}$ is a real vector space or, more precisely, a vector lattice and, hence, the abstract payoffs themselves, as elements of this space, are vectors. The nonnegative cone $\mathbf{\Pi}_+$ of $\mathbf{\Pi}$ contains precisely those abstract payoffs h that represent an amount of contingent wealth that is nonnegative no matter what is the state of the foreign exchange market. It is clear that for any abstract payoff h in $\mathbf{\Pi}_+$, the actual payoffs h_d and h_f have to be in turn members of the nonnegative cone Π_+ of Π, that is, the set of all nonnegative real-valued functions on the set of positive real numbers. Therefore, a consistency requirement is that the Kelvin transform, as the operator that maps h_d to h_f, and vice versa, must preserve the nonnegative cone Π_+ of Π. As it is easy to see, this is indeed the case.

We formalize the act of observing an abstract payoff via its actual payoff on either side of the foreign exchange market as a particular choice of a basis for the space of all abstract payoffs $\mathbf{\Pi}$. In the domestic side of the foreign exchange market, this basis, the domestic basis, is chosen in such a way that the coordinates, indexed by the exchange rate S, of the abstract payoff h in $\mathbf{\Pi}$ with respect to this basis are the values of an actual payoff $h_d(S)$. In this way the actual payoff h_d, as a function of the exchange rate S, is the coordinate description of the abstract payoff h in the domestic basis. In the foreign side of the foreign exchange market, this basis, the foreign basis, is chosen in such a way that the coordinates, indexed by the exchange rate S', of the abstract payoff h with respect to this basis are the values of an actual payoff $h_f(S')$. In this way the actual payoff h_f, as a function of the exchange rate S', is the coordinate description of the abstract payoff h in the foreign basis.

Since an actual payoff h_d or h_f is an arbitrary real-valued function defined on positive real numbers, that is, an element of the function space Π, the space $\mathbf{\Pi}$ of all abstract payoffs h and the function space Π are isomorphic as vector spaces. In fact, $\mathbf{\Pi}$ and Π are isomorphic as vector lattices, where the partial order on Π is generated by the nonnegative cone Π_+ and the lattice operations of supremum and infimum on Π are defined as the pointwise maximum and minimum.

In this formalism, the act of changing sides in the foreign exchange market is described by a linear change-of-basis operator that acts on the space Π, or more precisely, by an isomorphism of Π as a vector lattice. Due to the non-preferability of each side of the foreign exchange market, this operator has to perform both the change of basis from domestic to foreign and the change of basis from foreign to domestic. In addition, by performing the change of basis twice, starting with either side of the market, we have to restore the original basis. Therefore, the change-of-basis operator has to be a faithful representation of the cyclic group $\mathbb{Z}_2$ on Π.

In order to define explicitly the change-of-basis operator we note that, for any abstract payoff, this operator has to map the expression of this payoff in the domestic basis, that is, the domestic actual payoff, to the expression of this abstract payoff in the foreign basis, that is, to the foreign actual payoff, and vice versa. This means that for each abstract payoff h in $\mathbf{\Pi}$, the change-of-basis operator has to map the domestic actual payoff h_d in Π to the foreign actual payoff h_f in Π, and the foreign actual payoff h_f in Π to the domestic actual payoff h_d in Π. Therefore it is the Kelvin transform that is the change-of-basis operator as can be seen from the equalities in (1.3). As it is easy to verify, the Kelvin transform $\mathbf{K}$ as a linear operator on Π or, more precisely, an isomorphism of Π as a vector lattice, is indeed a faithful representation of the cyclic group $\mathbb{Z}_2$ on Π.

The $\mathbb{Z}_2$ symmetry in the foreign exchange market at the level of payoffs has its counterpart at the level of foreign exchange options.

Consider an economy without transaction costs in which trading is allowed at any time in a *trading time set* $\mathcal{T}$, defined as an arbitrary closed subset of the real numbers. Denote by $S_\tau > 0$ the price (of one unit) of the (only) underlying security at time τ in $\mathcal{T}$. Whenever no ambiguity is likely, we will write S in place of S_τ.

A *European option* (on the underlying security) with inception time t, expiration time T (t and T are in $\mathcal{T}$ with $t \leq T$), and with a payoff g in Π_+ is a contract that gives the right, but not the obligation, to receive the payoff $g(S_T)$ at expiration time T, where the price of the underlying security is S_T at this time T.

Among European options, one of the most widely traded are European call and put options. A European call and put options are European options with call and put payoffs $(\cdot - X)^+$ and $(X - \cdot)^+$, where $X > 0$ is the strike price. Financially, European call and put options give the right but not the obligation to by or to sell one unit of an underlying security at the strike price at the expiration time.

For each t and T in $\mathcal{T}$ with $t \leq T$, denote by $\mathbf{V}(t,T)$ the operator that maps a payoff of a European option with inception time t and expiration time T to the value of this option at inception time t as a function of the price S_t of the underlying security at this time t. Since the value of a European option at its expiration time is equal to its payoff, $\mathbf{V}(t,T)$ is the identity operator whenever inception time t is equal to expiration time T.

By definition, the domain of the operator $\mathbf{V}(t,T)$ is in Π_+. It turns out that, due to the no-free-lunch argument, the range of the operator $\mathbf{V}(t,T)$ is also in Π_+. Moreover, the operator $\mathbf{V}(t,T) : \Pi_+ \to \Pi_+$ admits a unique extension to an operator whose domain is a linear subspace of Π and whose range is in Π. Denote this new operator again by $\mathbf{V}(t,T)$. Due to the no-arbitrage argument, the operator $\mathbf{V}(t,T) : \Pi \to \Pi$ is a linear operator and, due to the previous discussion, it preserves the nonnegative cone Π_+ in Π. Moreover, due to the previous paragraph, $\mathbf{V}(t,t)$ is the identity operator on Π for each t in $\mathcal{T}$.

These are all the structural properties of the evolution operators $\mathbf{V}(t,T)$ that we need in order to present our symmetry in the foreign exchange market. For the complete analysis of the minimal set of such properties as dictated by financial reasons see [18].

We comment that there are important reasons for not imposing any further technical restrictions on the operators $\mathbf{V}(t,T)$, including their domains and ranges. First, as we indicated in the Preface, we do not wish to overload the book with too many fine details only of interest to pure mathematicians. Second, because these details might obscure the structure of the symmetry, which is purely algebraic in nature. Third, and much more importantly, we would like to be as independent as possible of assumptions on the particular nature of a stochastic process for the prices of the underlying security. Moreover, we would like to be free of the very assumption of the existence of such a stochastic process. How these stochastic issues might arise from such further technical restrictions on the operators $\mathbf{V}(t,T)$ we will see in Chapter 2.

The operators $\mathbf{V}(t,T)$, with t and T in $\mathcal{T}$ and $t \leq T$, contain all the information about the model of the market in which the European options are being priced. That is why we will say that a *market environment*, or simply *environment*, is given by, or simply is, the family of evolution operators

$$\mathbf{V} = \{\mathbf{V}(t,T) : \Pi \to \Pi \,|\, t, T \in \mathcal{T}, t \leq T\}.$$

In the foreign exchange market in which the underlying security is domestic or foreign currency, when we refer to European options in currency d on currency f or European options in currency f on currency d, we will replace the operator $\mathbf{V}(t,T)$ by $\mathbf{V}_d(t,T)$ or $\mathbf{V}_f(t,T)$ respectively. In this case the definitions of the operators $\mathbf{V}_d(t,T)$ and $\mathbf{V}_f(t,T)$ read:

$$\mathcal{E}_d(t,T,g_d) = \mathbf{V}_d(t,T)g_d \quad \text{and} \quad \mathcal{E}_f(t,T,g_f) = \mathbf{V}_f(t,T)g_f,$$

where $\mathcal{E}_d(t,T,g_d) = \mathcal{E}_d(t,T,g_d)(S_t)$ and $\mathcal{E}_f(t,T,g_f) = \mathcal{E}_f(t,T,g_f)(S'_t)$ denote the values at the inception time t of the European options in currency d on currency f and in currency f on currency d with expiration time T and with payoffs g_d and g_f.

In this notation, the values of the foreign exchange European call and put options can be represented as

$$C_d(t,T,S,X) = \mathcal{E}_d(t,T,(\cdot - X)^+)(S),$$
$$P_d(t,T,S,X) = \mathcal{E}_d(t,T,(X - \cdot)^+)(S),$$

and

$$C_f(t,T,S',X') = \mathcal{E}_f(t,T,(\cdot - X')^+)(S'),$$
$$P_f(t,T,S',X') = \mathcal{E}_f(t,T,(X' - \cdot)^+)(S').$$

As an illustration, the market environments corresponding to the binomial model of Cox, Ross, and Rubinstein [2] and to the Black and Scholes model [1] will be given explicitly in the foreign exchange setting and, following [18], will be called the binomial and the Black and Scholes market environments. Furthermore, we will show how the well known convergence, indicated originally in [2], of the values of European options in the binomial model to the values of these options in the Black and Scholes model can be expressed in terms of the convergence of the evolution operators in the binomial market environment to the corresponding evolution operators in the Black and Scholes market environment. This will imply that the symmetry relationships for foreign exchange options valued in the Black and Scholes market environment hold at all levels of their approximation in the binomial market environment. This, in turn, will be crucial in the application of our symmetry to software development.

In the same way that we have united financially equivalent payoffs on the opposite sides of the foreign exchange market into an abstract payoff, we can unite the financially equivalent foreign exchange options on the opposite sides of the market into an abstract option.

An *abstract European option* with inception time t, expiration time T (t and T are in $\mathcal{T}$, and $t \leq T$), and with an abstract payoff g in $\mathbf{\Pi}_+$ is a contract that gives the right, but not the obligation, to receive the abstract payoff g at the expiration time T.

For each t and T in $\mathcal{T}$ with $t \leq T$, denote by $\mathcal{V}(t,T)$ the operator that maps an abstract payoff g of an abstract European option with inception time t and expiration time T to the value of the option denoted by $\mathcal{E}(t,T,g)$ at the inception time t, that is

$$\mathcal{E}(t,T,g) = \mathcal{V}(t,T)\,g,$$

where g is in $\mathbf{\Pi}_+$. Since, by the no-arbitrage argument, the value of an abstract European option at its expiration time is equal to its abstract payoff, $\mathcal{V}(t,T)$ is the identity operator whenever inception time t is equal to expiration time T.

By definition, the domain of the operator $\mathcal{V}(t,T)$ is in $\mathbf{\Pi}_+$. Due to the no-free-lunch argument, the range of the operator $\mathcal{V}(t,T)$ is also in $\mathbf{\Pi}_+$. Moreover, the operator $\mathcal{V}(t,T) : \mathbf{\Pi}_+ \to \mathbf{\Pi}_+$ admits a unique extension to an operator whose domain is a linear subspace in $\mathbf{\Pi}$ and whose range is in $\mathbf{\Pi}$. Denote this new operator again by $\mathcal{V}(t,T)$. We call $\mathcal{V}(t,T)$ an *abstract evolution operator*. Due to the no-arbitrage argument, the operator $\mathcal{V}(t,T) : \mathbf{\Pi} \to \mathbf{\Pi}$ is a linear operator and, due to the previous discussion, it preserves the nonnegative cone $\mathbf{\Pi}_+$ in $\mathbf{\Pi}$. Moreover, due to the previous paragraph, $\mathcal{V}(t,t)$ is the identity operator on $\mathbf{\Pi}$ for each t in $\mathcal{T}$.

In the domestic basis for $\mathbf{\Pi}$ the abstract evolution operator $\mathcal{V}(t,T)$ is given by $\mathbf{V}_d(t,T)$ and in the foreign basis for $\mathbf{\Pi}$ the abstract evolution operator is given by $\mathbf{V}_f(t,T)$. Since $\mathbf{V}_d(t,T)$ and $\mathbf{V}_f(t,T)$ are the same operator $\mathcal{V}(t,T)$ expressed in different bases with the change-of-basis operator being the Kelvin transform $\mathbf{K}$, they are related by the similarity transform:

$$\begin{aligned} \mathbf{V}_d(t,T) &= \mathbf{K}\,\mathbf{V}_f(t,T)\,\mathbf{K}, \\ \mathbf{V}_f(t,T) &= \mathbf{K}\,\mathbf{V}_d(t,T)\,\mathbf{K}, \end{aligned}$$

(1.4)

where we made use of the fact that $\mathbf{K}^{-1} = \mathbf{K}$ since $\mathbf{K}$ is a faithful representation of the group $\mathbb{Z}_2$ on Π.

For any given model of foreign exchange option markets, these symmetry relationships present an additional requirement, a consistency requirement, and therefore they should be directly verified on a case-by-case basis. For example, as will be shown later in the book, in the two major cases, the binomial and the Black and Scholes market environments, this consistency requirement is satisfied.

The symmetry relationships (1.4) have the financial interpretation of a symmetry that associates financially equivalent European options on the opposite sides of the foreign exchange market.

In more detail, the expression of $\mathcal{E}(t,T,g)$, as an element of $\mathbf{\Pi}$, even of $\mathbf{\Pi}_+$, in the domestic basis is the value $\mathcal{E}_d(t,T,g_d)$ in Π_+ of the European option in currency d on currency f with inception time t, expiration time T and with payoff g_d, where g_d is the expression of the abstract payoff g in the domestic basis. On the other hand, the expression of $\mathcal{E}(t,T,g)$ in the foreign basis is the value $\mathcal{E}_f(t,T,g_f)$ in Π_+ of the European option in currency f on currency d with inception time t, expiration time T and with payoff g_f, where g_f is the expression of the abstract payoff g in the foreign basis.

Therefore the value $\mathcal{E}_f(t,T,g_f)$, converted to currency d, of the European option in currency f on currency d with payoff g_f has to be equal to the value $\mathcal{E}_d(t,T,g_d)$ of the European option in currency d on currency f with payoff g_d, where g_f is the payoff g_d converted to currency f. If this is not the case, then there is an arbitrage opportunity in the foreign exchange market. Similarly, the value $\mathcal{E}_d(t,T,g_d)$, converted to currency f has to be equal to the value $\mathcal{E}_f(t,T,g_f)$, where g_d is the payoff g_f converted to currency d. Now, for any g_d and g_f in Π_+ the relationships (1.4) read

$$\begin{aligned} \mathcal{E}_d(t,T,g_d) &= \mathbf{K}\,\mathcal{E}_f(t,T,\mathbf{K}\,g_d), \quad g_d \in \Pi_+, \\ \mathcal{E}_f(t,T,g_f) &= \mathbf{K}\,\mathcal{E}_d(t,T,\mathbf{K}\,g_f), \quad g_f \in \Pi_+. \end{aligned}$$

(1.5)

The analogous symmetry relationships are also valid for Bermudan and American foreign exchange options with general time-dependent payoffs allowing for introducing abstract Bermudan and American options as well. Furthermore, in the particular cases of European, Bermudan, and American foreign exchange call and put options these symmetry relationships take the form of the relationships in (1.1) and (1.2).

The concepts of an abstract payoff, an abstract evolution operator, and an abstract option allow for a coordinate-free, that is, a (domestic-foreign) basis-independent description of financial phenomena in a foreign exchange market.

The reason why we present the symmetry relationships for European, Bermudan and American options in the foreign exchange market with general payoffs is not just for the sake of complete generality. Even for the standard case of European, Bermudan and American call and put options, it is often advantageous to smooth the corners of their payoffs so that we are forced to consider payoffs of a general type. The practical reason for this smoothing is that the resulting options are easier to hedge when they are near the money and close to expiration. We will give an example of how the symmetry relationships can be used to guide the choice of the new smoothed payoffs so that the form of the symmetry relationships (1.1) and (1.2) for European, Bermudan, and American call and put options is maintained for the resulting options with the smoothed payoffs.

An even more important reason for the above generality is that it provides a framework which allows for the analysis of the symmetry relationships for one of the major classes of exotic foreign exchange options, namely barrier options. In turn, we consider these options in the generality of an arbitrary time-dependent barrier which can be activated not only on the entire life of the option, but on any discrete set of times during its life. Furthermore, we consider barrier options of European, Bermudan and American styles with the underlying European, Bermudan and American options having general and, in the latter two cases, time-dependent payoffs. As was shown in [18], these barrier options can be viewed as Bermudan options with certain payoffs for discretely activated barriers or as American options with certain payoffs for continuously activated barriers.

Finally, we establish the symmetry relationships for the Greek letters of foreign exchange options, in particular for delta and gamma, and the symmetry relationships for foreign exchange European, Bermudan, and American call and put options in the particular case of a market environment which we call the exchange-rate homogeneous market environment.

The symmetry relationships for foreign exchange options have many practical applications that will be presented in more detail in Chapter 6.

For example, the symmetry relationships provide a means for detecting a new type of true arbitrage in the foreign exchange option market directly from the market data.

They also provide a means of selecting, for a given foreign exchange portfolio, which may include options, a financially equivalent portfolio on the opposite side of a foreign exchange market, when for some reason it is not convenient or not possible to achieve the original portfolio. This is, for example, the case when one needs to take an option position and the foreign exchange option market is one-sided in the sense that the desired option is not listed on the official exchanges on the desired side of the market but its financially equivalent option is listed on such exchanges on the opposite side of the market.

At the same time, the symmetry relationships provide a screen to help in detecting inconsistent models of foreign exchange option markets.

Another major application is that the symmetry relationships allow one to instantly value and analyze the foreign counterpart, and its Greek letters, of a domestic foreign exchange option or portfolio of such options without lengthy additional re-calculations.

In practice, to apply one of the symmetry relationships for such instant valuation and analysis of a foreign counterpart to an option that does not have a closed-form solution, some approximation scheme has to be employed to first find the approximate value of this option itself. For example, even to compute the value in the Black and Scholes market environment of foreign exchange European options with an arbitrary payoff or of foreign exchange American call and put options, a standard way to proceed is to use the corresponding binomial model. Then, in order to find the approximate value of the foreign counterpart, one can apply the symmetry relationship to the approximate value of the option itself and still maintain control of the precision at all levels of approximation. The reason why this can be done is that the symmetry relationships are valid at all levels of the standard approximation schemes, namely approximation via binomial models or via Bermudan options. This allows, among other things, for the further improvement of the computational efficiency of modern high-speed binomial methods that achieve their speed by dramatically reducing the number of steps needed in the binomial model while still preserving the desired precision.

The symmetry relationships also have applications to the development and testing of algorithms and software to value and analyze portfolios of foreign exchange options, both for in-house and commercial use. In this regard, in addition to all the benefits described above, the symmetry relationships, as well as cutting memory usage in half, allow for a considerable reduction of coding and debugging time and, in turn, a reduction in the cost of software development.

The symmetry relationships can be used to guide the choice in smoothing the corners of payoffs of European, Bermudan, and American options such as European, Bermudan, and American call and put options. The practical reason for this smoothing is that the resulting options with smoothed payoffs are easier to hedge, for example, in the Black and Scholes market environment.

Finally, we note that the symmetry relationships presented in this book are not limited to foreign exchange markets and, in fact, remain valid for any financial markets with exchange of arbitrary underlying securities.

Part I

Financial Matters

Chapter 2

Market Environment

In this chapter we present the framework of a market environment in which foreign exchange options are being priced as a family of nonnegative evolution operators. This framework allows us to handle the symmetry relationships at the level of foreign exchange options without assumptions on the nature of a probability distribution for exchange rates, and, in fact, without the very assumption of the existence of such a distribution. The framework of a market environment was introduced by Kholodnyi in [18] and, both in this chapter and in subsequent chapters, we freely reproduce relevant concepts, terminology, and results first presented in that article. To illustrate this idea, we present explicitly two major market environments corresponding to the binomial model of Cox, Ross, and Rubinstein [2] and to the Black and Scholes model [1], and, following [18], we call them the binomial and the Black and Scholes market environments. Furthermore, we show how the well known convergence, indicated originally in [2], of the values of European options in the binomial model to the values of these options in the Black and Scholes model can be expressed in terms of the convergence of the evolution operators in the binomial market environment to the corresponding evolution operators in the Black and Scholes market environment. This will imply that the symmetry relationships for foreign exchange options valued in the Black and Scholes market environment hold at all levels of their approximation in the binomial market environment. This, in turn, will be crucial in the application of our symmetry to software development.

2.1   The Framework of a Market Environment

Consider an economy without transaction costs in which trading is allowed at any time in a *trading time set* $\mathcal{T}$, where $\mathcal{T}$ is an arbitrary closed subset of the real numbers. Denote by $S_\tau > 0$ the price (of one unit) of the (only) underlying security at time τ in $\mathcal{T}$. Whenever no ambiguity is likely, we will write S in place of S_τ. Denote by Π the vector space of all real-valued functions on the set of positive real numbers $\mathbb{R}_+$. Let Π_+ be the nonnegative cone of Π, that is, the set of all nonnegative real-valued functions on the set of positive real numbers.

A *European option* (on the underlying security) with inception time t, expiration time T (t and T are in $\mathcal{T}$ with $t \leq T$), and with a payoff g in Π_+ is a contract that gives the right, but not the obligation, to receive the payoff $g(S_T)$ at expiration time T, where the price of the underlying security is S_T at this time T.

Among European options, one of the most widely traded are European call and put options.

A *European call option* and a *European put option* are European options with payoffs $(\cdot - X)^+$ and $(X - \cdot)^+$ in Π_+, where $X > 0$ is called the *strike price*, and where

$$x^+ = \begin{cases} x & \text{if } x \geq 0, \\ 0 & \text{if } x < 0. \end{cases}$$

The payoffs $(\cdot - X)^+$ and $(X - \cdot)^+$ in Π_+ are called a *call payoff* and a *put payoff* respectively. Financially, European call and put options give the right but not the obligation to buy or to sell one unit of an underlying security at the strike price at the expiration time.

We comment that we are using the standard notation which distinguishes a function $y(\cdot)$ from its value $y(x)$ for the argument x.

For each t and T in $\mathcal{T}$ with $t \leq T$, denote by $\mathbf{V}(t, T)$ the operator that maps a payoff of a European option with inception time t and expiration time T to the value of this option at inception time t as a function of the price S_t of the underlying security at this time t. Since the value of a European option at its expiration time is equal to its payoff, $\mathbf{V}(t, T)$ is the identity operator whenever inception time t is equal to expiration time T.

By definition, the domain of the operator $\mathbf{V}(t, T)$ is in Π_+. Due to the no-free-lunch argument, the range of the operator $\mathbf{V}(t, T)$ is also in Π_+. Moreover, the operator $\mathbf{V}(t, T) : \Pi_+ \to \Pi_+$ admits a unique extension to an operator whose domain is a linear subspace of Π and whose range is in Π. Denote this new operator again by $\mathbf{V}(t, T)$. Due to the no-arbitrage argument, the operator $\mathbf{V}(t, T) : \Pi \to \Pi$ is a linear operator and, due to the previous discussion, it preserves the nonnegative cone Π_+ in Π. Moreover, due to the previous paragraph, $\mathbf{V}(t, t)$ is the identity operator on Π for each t in $\mathcal{T}$.

These are all the structural properties of the evolution operators $\mathbf{V}(t, T)$ that we need in order to present our symmetry in the foreign exchange market. For the complete analysis of the minimal set of such properties as dictated by financial reasons see [18].

We comment that there are important reasons for not imposing any further technical restrictions on the operators $\mathbf{V}(t, T)$ such as their domains and ranges. First, as we indicated in the Preface and Introduction, we do not wish to overload the book with too many fine details only of interest to pure mathematicians. Second, because these details might obscure the structure of the symmetry, which is purely algebraic in nature. Third, and much more importantly, we want to be as independent as possible of assumptions on the particular nature of a stochastic process for the prices of the underlying security. Moreover, we would like to be free of the very assumption of the existence of such a stochastic process. One way in which these stochastic issues might arise from such further technical restrictions is as follows. The kernel $\mathbf{V}(t, T)(S_t, S_T)$ of the operator $\mathbf{V}(t, T)$, as a function of S_t, is financially the value at inception time t of the European option with expiration time T and with the Dirac delta function $\delta(\cdot - S_T)$ as its payoff, that is, the value of the Arrow-Debreu security understood in the generalized sense. If a model of the market place is chosen that treats the dynamics of the prices of the underlying security as a stochastic process, then this kernel is financially the discounted transition probability distribution of the prices of the underlying security. Therefore, any such further restrictions on the operator $\mathbf{V}(t, T)$ are effectively assumptions on the particular nature of the stochastic process for the prices of the underlying security.

The operators $\mathbf{V}(t, T)$, with t and T in $\mathcal{T}$ and $t \le T$, contain all the information about the model of the market in which the European options are being priced. That is why we will say that a *market environment*, or simply *environment*, is given by, or simply is, the family of evolution operators

$$\mathbf{V} = \{\mathbf{V}(t, T) : \Pi \to \Pi \,|\, t, T \in \mathcal{T}, t \le T\}.$$

The *time-homogeneous market environment* is a market environment $\mathbf{V}$ in which the evolution operators $\mathbf{V}(t, T)$ are functions only of $T - t$. If a model of the market place is chosen that treats the dynamics of the prices of the underlying security as a stochastic process, then a time-homogeneous market environment corresponds to a time-homogeneous stochastic process.

In the foreign exchange market in which the underlying security is the domestic or foreign currency, when we refer to European options in currency d on currency f or European options in currency f on currency d, we will replace the operator $\mathbf{V}(t,T)$ by $\mathbf{V}_d(t,T)$ or $\mathbf{V}_f(t,T)$ respectively. In this case the definitions of the operators $\mathbf{V}_d(t,T)$ or $\mathbf{V}_f(t,T)$ read:

$$\mathcal{E}_d(t,T,g_d) = \mathbf{V}_d(t,T)g_d \quad \text{and} \quad \mathcal{E}_f(t,T,g_f) = \mathbf{V}_f(t,T)g_f,$$

where $\mathcal{E}_d(t,T,g_d) = \mathcal{E}_d(t,T,g_d)(S_t)$ and $\mathcal{E}_f(t,T,g_f) = \mathcal{E}_f(t,T,g_f)(S_t')$ denote the values at the inception time t of the European options in currency d on currency f and in currency f on currency d with expiration time T and with payoffs g_d and g_f.

In this notation, the values of the foreign exchange European call and put options can be represented as

$$C_d(t,T,S,X) = \mathcal{E}_d(t,T,(\cdot - X)^+)(S),$$
$$P_d(t,T,S,X) = \mathcal{E}_d(t,T,(X - \cdot)^+)(S),$$

and

$$C_f(t,T,S',X') = \mathcal{E}_f(t,T,(\cdot - X')^+)(S'),$$
$$P_f(t,T,S',X') = \mathcal{E}_f(t,T,(X' - \cdot)^+)(S'),$$

where

$$(\cdot - X)^+ \quad \text{and} \quad (X - \cdot)^+$$

are the call and put payoffs in currency d with strike price X, and

$$(\cdot - X')^+ \quad \text{and} \quad (X' - \cdot)^+$$

are the call and put payoffs in currency f with strike price X'.

In this way, for the foreign exchange market we have two market environments $\mathbf{V}_d$ and $\mathbf{V}_f$, one for each side of the market, domestic and foreign:

$$\mathbf{V}_d = \{\mathbf{V}_d(t,T) : \Pi \to \Pi \,|\, t,T \in \mathcal{T}, t \leq T\}$$

and

$$\mathbf{V}_f = \{\mathbf{V}_f(t,T) : \Pi \to \Pi \,|\, t,T \in \mathcal{T}, t \leq T\}.$$

However, in the next Chapter we will see that the market environments $\mathbf{V}_d$ and $\mathbf{V}_f$ are not independent but, in fact, are related via the symmetry relationships for the evolution operators $\mathbf{V}_d(t,T)$ and $\mathbf{V}_f(t,T)$.

To illustrate the framework of the market environment, in the next two sections we give explicitly in the foreign exchange setting the evolution operators of the two major market environments, namely the binomial and Black and Scholes market environments corresponding to the binomial model of Cox, Ross, and Rubinstein [2] and to the Black and Scholes model [1].

2.2   The Binomial Market Environment

Consider a foreign exchange market in which trading is allowed only at times t_l, $l = 0, 1, ..., n$, such that $0 < t_0 < t_1 < ... < t_{n-1} < t_n$, that is, with the trading time set $\mathcal{T} = \{t_l : l = 0, 1, ..., n\}$.

In this section we present the market environments $\mathbf{V}_d^B$ and $\mathbf{V}_f^B$ with evolution operators $\mathbf{V}_d^B(t, T) : \Pi \to \Pi$ and $\mathbf{V}_f^B(t, T) : \Pi \to \Pi$ with t and T in $\mathcal{T}$ and $t \leq T$ corresponding to the binomial model of Cox, Ross, and Rubinstein [2]. In fact, we present these market environments with the evolution operators corresponding to a straightforward extension of the binomial model of Cox, Ross, and Rubinstein in which bond is replaced by a general risky security. These market environments were introduced in [18], and, following this article, we call them the *binomial market environments*. Following this article, we also present in which sense the binomial market environments $\mathbf{V}_d^B$ and $\mathbf{V}_f^B$ with the evolution operators $\mathbf{V}_d^B(t, T)$ and $\mathbf{V}_f^B(t, T)$ defined on Π, the vector space of all real-valued functions on $\mathbb{R}_+$, correspond to the binomial model of Cox, Ross, and Rubinstein based on a binomial tree.

Suppose that at time t_l with $l = 0, 1, \ldots, n-1$ in the trading time set $\mathcal{T}$ the foreign exchange market is in state S as expressed by the value of the exchange rate between currency d and currency f. Then at time t_{l+1} the market can move to either one of two states $u_l(S)S$ or $v_l(S)S$ as expressed by the values of the exchange rate between currency d and currency f at time t_{l+1}. Due to the positivity of the exchange rates, the multipliers $u_l(S)$ and $v_l(S)$ are positive.

We assume that none of these states of the foreign exchange market are a priori excluded.

On the other hand, viewed from the opposite side of the foreign exchange market, at time t_l the market is in state $S' = 1/S$ as expressed by the value of the exchange rate between currency f and currency d. Then at time t_{l+1} the market can move to either one of two states $u_l'(S')S'$ or $v_l'(S')S'$ as expressed by the values of the exchange rate between currency f and currency d at time t_{l+1}. Clearly, $u_l'(S') = 1/u_l(1/S')$ and $v_l'(S') = 1/v_l(1/S')$.

Suppose that in the domestic side of the foreign exchange market, in addition to the foreign currency, we have another security with a positive return per unit of currency d depending upon the state of the market as follows. If the market moves from the state S at time t_l to the state $u_l(S)S$ at time t_{l+1}, the return on this security is $\rho_{u,l}(S)$, while if the market moves to the state $v_l(S)S$, the return is $\rho_{v,l}(S)$. Similarly, suppose that in the foreign side of the foreign exchange market, in addition to the domestic currency, we have another security with a positive return per unit of currency f depending upon the state of the market as follows. If the market moves from the state S' at time t_l to the state $u'_l(S')S'$ at time t_{l+1}, the return on this security is $\rho'_{u,l}(S')$, while if the market moves to the state $v'_l(S')S'$, the return is $\rho'_{v,l}(S')$.

Note that in the particular case where the returns on these additional securities do not depend on the future state of the market, that is, if $\rho_{u,l}(S) = \rho_{v,l}(S)$ and $\rho'_{u,l}(S') = \rho'_{v,l}(S')$ for all S and S', the securities are riskless. In this case, we can interpret these securities as domestic and foreign pure discount bonds with face values of one unit of currencies d and f, with maturity time t_{l+1}, and with values $1/\rho_{u,l}(S)$ and $1/\rho'_{u,l}(S')$ at their inception time t_l.

Returning to the general case, we assume that for each time t_l and for each state of the foreign exchange market at time t_l, the market is complete and arbitrage-free, that is,

$$(2.1) \qquad v_l(S)\frac{\rho'_{v,l}(1/S)}{\rho_{v,l}(S)} \;\lessgtr\; 1 \;\lessgtr\; u_l(S)\frac{\rho'_{u,l}(1/S)}{\rho_{u,l}(S)}$$

$$(2.2) \qquad \frac{1}{u_l(1/S')}\frac{\rho_{u,l}(1/S')}{\rho'_{u,l}(S')} \;\lessgtr\; 1 \;\lessgtr\; \frac{1}{v_l(1/S')}\frac{\rho_{v,l}(1/S')}{\rho'_{v,l}(S')}.$$

We note that the first and second alternatives in inequalities (2.1) are equivalent to the first and second alternatives in inequalities (2.2), respectively.

Denote by $p_{u,l}(S)$ and $p_{v,l}(S)$ the risk-neutral transition probabilities for the foreign exchange market to move to the state $u_l(S)S$ or $v_l(S)S$ at time t_{l+1} given that the market was in the state S at time t_l. From the opposite side of the foreign exchange market, denote by $p'_{u,l}(S')$ and $p'_{v,l}(S')$ the risk-neutral transition probabilities for the market to move to the state $u'_l(S')S'$ or $v'_l(S')S'$ at time t_{l+1} given that the market was in the state S' at time t_l. By a straightforward extension of the binomial model of Cox, Ross and Rubinstein in which the bond is replaced by a general risky security, these risk-neutral transition probabilities are given by

$$p_{u,l}(S) = \frac{\frac{\rho_{v,l}(S)}{\rho'_{v,l}(1/S)} - v_l(S)}{\frac{\rho'_{u,l}(1/S)\rho_{v,l}(S)}{\rho'_{v,l}(1/S)\rho_{u,l}(S)}u_l(S) - v_l(S)},$$

(2.3)

$$p_{v,l}(S) = \frac{u_l(S) - \frac{\rho_{u,l}(S)}{\rho'_{u,l}(1/S)}}{u_l(S) - \frac{\rho_{u,l}(S)\rho'_{v,l}(1/S)}{\rho_{v,l}(S)\rho'_{u,l}(1/S)}v_l(S)},$$

$$p'_{u,l}(S') = \frac{\frac{\rho'_{v,l}(S')}{\rho_{v,l}(1/S')} - \frac{1}{v_l(1/S')}}{\frac{\rho_{u,l}(1/S')\rho'_{v,l}(S')}{\rho_{v,l}(1/S')\rho'_{u,l}(S')}\frac{1}{u_l(1/S')} - \frac{1}{v_l(1/S')}},$$

(2.4)

$$p'_{v,l}(S') = \frac{\frac{1}{u_l(1/S')} - \frac{\rho'_{u,l}(S')}{\rho_{u,l}(1/S')}}{\frac{1}{u_l(1/S')} - \frac{\rho'_{u,l}(S')\rho_{v,l}(1/S')}{\rho'_{v,l}(S')\rho_{u,l}(1/S')}\frac{1}{v_l(1/S')}}.$$

It is easy to see that for all S and S'

$$0 < p_{u,l}(S), \ p_{v,l}(S) < 1 \ \ \text{and} \ \ p_{u,l}(S) + p_{v,l}(S) = 1,$$

and

$$0 < p'_{u,l}(S'), \ p'_{v,l}(S') < 1 \ \ \text{and} \ \ p'_{u,l}(S') + p'_{v,l}(S') = 1.$$

The desired evolution operators $\mathbf{V}_d^B(t,T)$ and $\mathbf{V}_f^B(t,T)$ are now defined by

(2.5)
$$\mathbf{V}_d^B(t,T) =$$
$$\mathbf{V}_d^B(t = t_i, t_{i+1})\mathbf{V}_d^B(t_{i+1}, t_{i+2})\ldots\mathbf{V}_d^B(t_{j-1}, t_j = T), \quad t, T \in \mathcal{T}, \, t < T,$$
$$\mathbf{V}_f^B(t,T) =$$
$$\mathbf{V}_f^B(t = t_i, t_{i+1})\mathbf{V}_f^B(t_{i+1}, t_{i+2})\ldots\mathbf{V}_f^B(t_{j-1}, t_j = T), \quad t, T \in \mathcal{T}, \, t < T,$$

where
(2.6)
$$(\mathbf{V}_d^B(t_l, t_{l+1})h_d)(S) =$$
$$\frac{p_{u,l}(S)}{\rho_{u,l}(S)}h_d(u_l(S)\,S) + \frac{p_{v,l}(S)}{\rho_{v,l}(S)}h_d(v_l(S)\,S), \quad h_d \in \Pi,$$

$$(\mathbf{V}_f^B(t_l, t_{l+1})h_f)(S') =$$
$$\frac{p'_{u,l}(S')}{\rho'_{u,l}(S')}h_f(u'_l(S')S') + \frac{p'_{v,l}(S')}{\rho'_{v,l}(S')}h_f(v'_l(S')S'), \quad h_f \in \Pi.$$

We comment that due to the definition of the evolution operators in a market environment $\mathbf{V}_d^B(t,t)$ and $\mathbf{V}_f^B(t,t)$ must be the identity operators on Π for each t in $\mathcal{T}$.

Now we show that the definition of the operators $\mathbf{V}_d^B(t,T)$ and $\mathbf{V}_f^B(t,T)$ is consistent with a straightforward extension of the binomial model of Cox, Ross and Rubinstein in which the bond is replaced by a general risky security.

Suppose that the initial state at time t_0 of the foreign exchange market is S as expressed by the value of the exchange rate between currency d and currency f at time t_0. Suppose that for any given state at time t_l, with $l = 0, 1, \ldots, n-1$, the market can move to either one of two states at time t_{l+1} so that at each time t_l in the trading time set $\mathcal{T}$, the market can be in one of 2^l states. We label these states at time t_l by pairs (l, m), where $m = 1, \ldots, 2^l$, in such a way that if at time t_l the market is in state (l, m), then at time t_{l+1} it can move either to the state $(l+1, 2m)$ or the state $(l+1, 2m-1)$.

In each of these states (l, m) of the foreign exchange market, denote the exchange rate between currency d and currency f by S_{lm}. If the market moves from the state (l, m) to the state $(l+1, 2m)$, the exchange rate S_{lm} moves to $S_{l+1,2m} = \hat{u}_l(S_{lm})S_{lm}$, while if the market moves to the state $(l+1, 2m-1)$, the exchange rate moves to $S_{l+1,2m-1} = \hat{v}_l(S_{lm})S_{lm}$. Since the exchange rates are positive, so are the multipliers $\hat{u}_l(S_{lm})$ and $\hat{v}_l(S_{lm})$.

We assume that none of these states of the foreign exchange market are a priori excluded.

For each of the states (l, m), the exchange rate viewed from the opposite side of the foreign exchange market, that is, the exchange rate between currency f and currency d, is given by $S'_{lm} = 1/S_{lm}$. If the market moves from the state (l, m) to the state $(l+1, 2m)$, the exchange rate S'_{lm} moves to $S'_{l+1,2m} = \hat{u}'_l(S'_{lm})S'_{lm}$, while if the market moves to the state $(l+1, 2m-1)$, the exchange rate moves to $S'_{l+1,2m-1} = \hat{v}'_l(S'_{lm})S'_{lm}$, where $\hat{u}'_l(S'_{lm}) = 1/\hat{u}_l(1/S'_{lm})$ and $\hat{v}'_l(S'_{lm}) = 1/\hat{v}_l(1/S'_{lm})$.

We call the grid

$$\Gamma_d^n = \Gamma_d^n\left((t_l, S_{lm}) : m = 1, \ldots, 2^l, l = 0, \ldots, n\right),$$

with $S_{01} = S$, the n-step binomial tree in currency d on currency f. Similarly, we call the corresponding grid

$$\Gamma_f^n = \Gamma_f^n\left((t_l, S'_{lm}) : m = 1, \ldots, 2^l, l = 0, \ldots, n\right),$$

with $S'_{01} = S'$, the n-step binomial tree in currency f on currency d. Figure 2.1 displays the binomial tree Γ_d^n with $n = 2$.

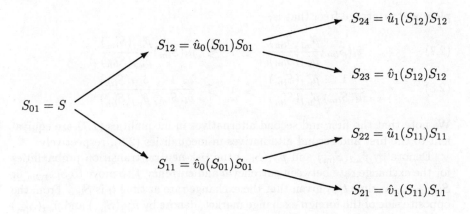

Figure 2.1: **The Binomial Tree Γ_d^n with $n = 2$.**

Suppose that in the domestic side of the foreign exchange market, in addition to the foreign currency, we have another security with a positive return per unit of currency d depending upon the state of the market as follows. If the market moves from the state (l, m) at time t_l to the state $(l + 1, 2m)$ at time t_{l+1}, the return on this security is $\hat{\rho}_{u,l}(S_{lm})$, while if the market moves to the state $(l + 1, 2m - 1)$, the return is $\hat{\rho}_{v,l}(S_{lm})$. Similarly, suppose that in the foreign side of the foreign exchange market, in addition to the domestic currency, we have another security with a positive return per unit of currency f depending upon the state of the market as follows. If the market moves from the state (l, m) at time t_l to the state $(l + 1, 2m)$ at time t_{l+1}, the return on this security is $\hat{\rho}'_{u,l}(S'_{lm})$, while if the market moves to the state $(l + 1, 2m - 1)$, the return is $\hat{\rho}'_{v,l}(S'_{lm})$.

Note that in the particular case where the returns on these additional securities do not depend on the future state of the market, that is, if $\hat{\rho}_{u,l}(S_{lm}) = \hat{\rho}_{v,l}(S_{lm})$ and $\hat{\rho}'_{u,l}(S'_{lm}) = \hat{\rho}'_{v,l}(S'_{lm})$ for all states (l, m), these securities are riskless. In this case, we can interpret the securities as domestic and foreign pure discount bonds with face values of one unit of currencies d and f, with maturity time t_{l+1}, and with values $1/\hat{\rho}_{u,l}(S_{lm})$ and $1/\hat{\rho}'_{u,l}(S'_{lm})$ at their inception time t_l.

We assume that for each state (l, m) the foreign exchange market is complete and arbitrage free, that is,

$$(2.7) \qquad \hat{v}_l(S_{lm}) \frac{\hat{\rho}'_{v,l}(S'_{lm})}{\hat{\rho}_{v,l}(S_{lm})} \lessgtr 1 \lessgtr \hat{u}_l(S_{lm}) \frac{\hat{\rho}'_{u,l}(S'_{lm})}{\hat{\rho}_{u,l}(S_{lm})}$$

$$(2.8) \qquad \frac{1}{\hat{u}_l(S_{lm})} \frac{\hat{\rho}_{u,l}(S_{lm})}{\hat{\rho}'_{u,l}(S'_{lm})} \lessgtr 1 \lessgtr \frac{1}{\hat{v}_l(S_{lm})} \frac{\hat{\rho}_{v,l}(S_{lm})}{\hat{\rho}'_{v,l}(S'_{lm})}.$$

We note that the first and second alternatives in inequalities (2.7) are equivalent to the first and second alternatives in inequalities (2.8), respectively.

Denote by $\hat{p}_{u,l}(S_{lm})$ and $\hat{p}_{v,l}(S_{lm})$ the risk-neutral transition probabilities for the exchange rate between currency d and currency f to move to $S_{l+1,2m}$ or $S_{l+1,2m-1}$ at time t_{l+1} given that the exchange rate at time t_l is S_{lm}. From the opposite side of the foreign exchange market, denote by $\hat{p}'_{u,l}(S'_{lm})$ and $\hat{p}'_{v,l}(S'_{lm})$ the risk-neutral transition probabilities for the exchange rate between currency f and currency d to move to $S'_{l+1,2m}$ or $S'_{l+1,2m-1}$ at time t_{l+1} given that the exchange rate at time t_l is S'_{lm}. By a straightforward extension of the binomial model of Cox, Ross and Rubinstein in which the bond is replaced by a general risky security, these risk-neutral transition probabilities are given by

$$(2.9) \qquad \hat{p}_{u,l}(S_{lm}) = \frac{\frac{\hat{\rho}_{v,l}(S_{lm})}{\hat{\rho}'_{v,l}(S'_{lm})} - \hat{v}_l(S_{lm})}{\frac{\hat{\rho}'_{u,l}(S'_{lm})\hat{\rho}_{v,l}(S_{lm})}{\hat{\rho}'_{v,l}(S'_{lm})\hat{\rho}_{u,l}(S_{lm})} \hat{u}_l(S_{lm}) - \hat{v}_l(S_{lm})},$$

$$\hat{p}_{v,l}(S_{lm}) = \frac{\hat{u}_l(S_{lm}) - \frac{\hat{\rho}_{u,l}(S_{lm})}{\hat{\rho}'_{u,l}(S'_{lm})}}{\hat{u}_l(S_{lm}) - \frac{\hat{\rho}_{u,l}(S_{lm})\hat{\rho}'_{v,l}(S'_{lm})}{\hat{\rho}_{v,l}(S_{lm})\hat{\rho}'_{u,l}(S'_{lm})} \hat{v}_l(S_{lm})},$$

$$(2.10) \qquad \hat{p}'_{u,l}(S'_{lm}) = \frac{\frac{\hat{\rho}'_{u,l}(S'_{lm})}{\hat{\rho}_{v,l}(S_{lm})} - \frac{1}{\hat{v}_l(S_{lm})}}{\frac{\hat{\rho}_{u,l}(S_{lm})\hat{\rho}'_{v,l}(S'_{lm})}{\hat{\rho}_{v,l}(S_{lm})\hat{\rho}'_{u,l}(S'_{lm})} \frac{1}{\hat{u}_l(S_{lm})} - \frac{1}{\hat{v}_l(S_{lm})}},$$

$$\hat{p}'_{v,l}(S'_{lm}) = \frac{\frac{1}{\hat{u}_l(S_{lm})} - \frac{\hat{\rho}'_{u,l}(S'_{lm})}{\hat{\rho}_{u,l}(S_{lm})}}{\frac{1}{\hat{u}_l(S_{lm})} - \frac{\hat{\rho}'_{u,l}(S'_{lm})\hat{\rho}_{v,l}(S_{lm})}{\hat{\rho}'_{v,l}(S'_{lm})\hat{\rho}_{u,l}(S_{lm})} \frac{1}{\hat{v}_l(S_{lm})}}.$$

It is easy to see that for all S_{lm} and S'_{lm}

$$0 < \hat{p}_{u,l}(S_{lm}), \ \hat{p}_{v,l}(S_{lm}) < 1 \quad \text{and} \quad \hat{p}_{u,l}(S_{lm}) + \hat{p}_{v,l}(S_{lm}) = 1,$$

and

$$0 < \hat{p}'_{u,l}(S'_{lm}), \ \hat{p}'_{v,l}(S'_{lm}) < 1 \quad \text{and} \quad \hat{p}'_{u,l}(S'_{lm}) + \hat{p}'_{v,l}(S'_{lm}) = 1.$$

Consider now the vector spaces of all real-valued functions $\Pi_d(t)$ and $\Pi_f(t)$, with t in $\mathcal{T}$, defined on the sections

$$\Gamma_d^n(t) = \{S_{lm} : m = 1, \ldots, 2^l, t = t_l\}$$

and

$$\Gamma_f^n(t) = \{S'_{lm} : m = 1, \ldots, 2^l, t = t_l\}$$

of the binomial trees Γ_d^n and Γ_f^n at the time t.

We note that the functions $\hat{u}_l = \hat{u}_l(S_{lm})$, $\hat{v}_l = \hat{v}_l(S_{lm})$, $\hat{\rho}_{u,l} = \hat{\rho}_{u,l}(S_{lm})$, $\hat{\rho}_{v,l} = \hat{\rho}_{v,l}(S_{lm})$, $\hat{p}_{u,l} = \hat{p}_{u,l}(S_{lm})$, and $\hat{p}_{v,l} = \hat{p}_{v,l}(S_{lm})$ are members of $\Pi_d(t_l)$. Similarly, we note that the functions $\hat{u}'_l = \hat{u}'_l(S'_{lm})$, $\hat{v}'_l = \hat{v}'_l(S'_{lm})$, $\hat{\rho}'_{u,l} = \hat{\rho}'_{u,l}(S'_{lm})$, $\hat{\rho}'_{v,l} = \hat{\rho}'_{v,l}(S'_{lm})$, $\hat{p}'_{u,l} = \hat{p}'_{u,l}(S'_{lm})$, and $\hat{p}'_{v,l} = \hat{p}'_{v,l}(S'_{lm})$ are members of $\Pi_f(t_l)$.

Denote by $Q_d(t) : \Pi \to \Pi_d(t)$ and by $Q_f(t) : \Pi \to \Pi_f(t)$ the operators that restrict the functions in Π defined on $\mathbb{R}_+$ to the functions in $\Pi_d(t)$ and $\Pi_f(t)$ defined on $\Gamma_d^n(t)$ and $\Gamma_f^n(t)$.

Now we are ready to show the promised consistency of the definition of the operators $\mathbf{V}_d^B(t, T)$ and $\mathbf{V}_f^B(t, T)$ with a straightforward extension of the binomial model of Cox, Ross and Rubinstein in which the bond is replaced by a general risky security.

We make the further assumption on the functions $u_l = u_l(S)$ and $v_l = v_l(S)$ that their restrictions to the section $\Gamma_d^n(t_l)$ of the binomial tree Γ_d^n satisfy

$$(2.11) \qquad Q_d(t_l)u_l = \hat{u}_l, \quad Q_d(t_l)v_l = \hat{v}_l.$$

This implies that the restrictions of the functions $u'_l = u'_l(S')$ and $v'_l = v'_l(S')$ to the section $\Gamma_f^n(t_l)$ of the binomial tree Γ_f^n satisfy

$$(2.12) \qquad Q_f(t_l)u'_l = \hat{u}'_l, \quad Q_f(t_l)v'_l = \hat{v}'_l.$$

We also make the further assumption on the functions $\rho_{u,l} = \rho_{u,l}(S)$ and $\rho_{v,l} = \rho_{v,l}(S)$ that their restrictions to the section $\Gamma_d^n(t_l)$ of the binomial tree Γ_d^n satisfy

$$(2.13) \qquad Q_d(t_l)\rho_{u,l} = \hat{\rho}_{u,l}, \quad Q_d(t_l)\rho_{v,l} = \hat{\rho}_{v,l}.$$

Similarly, we also make the further assumption on the functions $\rho'_{u,l} = \rho'_{u,l}(S')$ and $\rho'_{v,l} = \rho'_{v,l}(S')$ that their restrictions to the section $\Gamma_f^n(t_l)$ of the binomial tree Γ_f^n satisfy

$$(2.14) \qquad Q_f(t_l)\rho'_{u,l} = \hat{\rho}'_{u,l}, \quad Q_f(t_l)\rho'_{v,l} = \hat{\rho}'_{v,l}.$$

The restriction requirements (2.11)–(2.14) in combination with the relationships (2.3) and (2.9) imply that the restrictions of the functions $p_{u,l} = p_{u,l}(S)$ and $p_{v,l} = p_{v,l}(S)$ to the section $\Gamma_d^n(t_l)$ of the binomial tree Γ_d^n satisfy

$$(2.15) \qquad Q_d(t_l)p_{u,l} = \hat{p}_{u,l}, \quad Q_d(t_l)p_{v,l} = \hat{p}_{v,l}.$$

Similarly, the restriction requirements (2.11)–(2.14) in combination with the relationships (2.4) and (2.10) imply that the restrictions of the functions $p'_{u,l} = p'_{u,l}(S')$ and $p'_{v,l} = p'_{v,l}(S')$ to the section $\Gamma_f^n(t_l)$ of the binomial tree Γ_f^n satisfy

$$(2.16) \qquad Q_f(t_l)p'_{u,l} = \hat{p}'_{u,l}, \quad Q_f(t_l)p'_{v,l} = \hat{p}'_{v,l}.$$

The operators $\mathbf{V}_d^B(t,T) : \Pi \to \Pi$ and $\mathbf{V}_f^B(t,T) : \Pi \to \Pi$ with parameters defined according to the preceding restriction requirements, can be restricted to functions defined only on the binomial trees Γ_d^n and Γ_f^n. More precisely, for each t and T in $\mathcal{T}$ with $t \leq T$, there are linear operators $\hat{\mathbf{V}}_d^B(t,T) : \Pi_d(T) \to \Pi_d(t)$ and $\hat{\mathbf{V}}_f^B(t,T) : \Pi_f(T) \to \Pi_f(t)$, that we call the *restrictions* of the operators $\mathbf{V}_d^B(t,T)$ and $\mathbf{V}_f^B(t,T)$ to the binomial trees Γ_d^n and Γ_f^n, such that

$$(2.17) \qquad \begin{aligned} Q_d(t)\mathbf{V}_d^B(t,T) &= \hat{\mathbf{V}}_d^B(t,T)Q_d(T), \\ Q_f(t)\mathbf{V}_f^B(t,T) &= \hat{\mathbf{V}}_f^B(t,T)Q_f(T). \end{aligned}$$

Following [18] we call the families

$$\hat{\mathbf{V}}_d^B = \{\hat{\mathbf{V}}_d^B(t,T) : \Pi_d(T) \to \Pi_d(t) \,|\, t, T \in \mathcal{T}, t \leq T\}$$

and

$$\hat{\mathbf{V}}_f^B = \{\hat{\mathbf{V}}_f^B(t,T) : \Pi_f(T) \to \Pi_f(t) \,|\, t, T \in \mathcal{T}, t \leq T\}$$

of these restricted evolution operators $\hat{\mathbf{V}}_d^B(t,T)$ and $\hat{\mathbf{V}}_f^B(t,T)$ *restricted binomial market environments*.

These restricted operators $\hat{\mathbf{V}}_d^B(t,T)$ and $\hat{\mathbf{V}}_f^B(t,T)$ are explicitly given by

$$(2.18)$$
$$\hat{\mathbf{V}}_d^B(t,T) =$$
$$\hat{\mathbf{V}}_d^B(t = t_i, t_{i+1})\hat{\mathbf{V}}_d^B(t_{i+1}, t_{i+2})\ldots\hat{\mathbf{V}}_d^B(t_{j-1}, t_j = T), \quad t, T \in \mathcal{T}, \, t < T,$$
$$\hat{\mathbf{V}}_f^B(t,T) =$$
$$\hat{\mathbf{V}}_f^B(t = t_i, t_{i+1})\hat{\mathbf{V}}_f^B(t_{i+1}, t_{i+2})\ldots\hat{\mathbf{V}}_f^B(t_{j-1}, t_j = T), \quad t, T \in \mathcal{T}, \, t < T,$$

where

(2.19)
$$(\hat{\mathbf{V}}_d^B(t_l, t_{l+1})\hat{h}_d)(S_{lm}) =$$
$$\frac{\hat{p}_{u,l}(S_{lm})}{\hat{\rho}_{u,l}(S_{lm})}\hat{h}_d\left(\hat{u}_l(S_{lm})S_{lm}\right) + \frac{\hat{p}_{v,l}(S_{lm})}{\hat{\rho}_{v,l}(S_{lm})}\hat{h}_d\left(\hat{v}_l(S_{lm})S_{lm}\right),$$
$$(\hat{\mathbf{V}}_f^B(t_l, t_{l+1})\hat{h}_f)(S'_{lm}) =$$
$$\frac{\hat{p}'_{u,l}(S'_{lm})}{\hat{\rho}'_{u,l}(S'_{lm})}\hat{h}_f\left(\hat{u}'_l(S'_{lm})S'_{lm}\right) + \frac{\hat{p}'_{v,l}(S'_{lm})}{\hat{\rho}'_{v,l}(S'_{lm})}\hat{h}_f\left(\hat{v}'_l(S'_{lm})S'_{lm}\right),$$

with $\hat{h}_d$ in $\Pi_d(t_{l+1})$ and $\hat{h}_f$ in $\Pi_f(t_{l+1})$. It is clear that $\hat{\mathbf{V}}_d^B(t,t)$ and $\hat{\mathbf{V}}_f^B(t,t)$ are the identity operators on $\Pi_d(t)$ and $\Pi_f(t)$ for each t in $\mathcal{T}$.

These restrictions $\hat{\mathbf{V}}_d^B(t,T)$ and $\hat{\mathbf{V}}_f^B(t,T)$ of the operators $\mathbf{V}_d^B(t,T)$ and $\mathbf{V}_f^B(t,T)$ to the binomial trees Γ_d^n and Γ_f^n determine the values of European options in a straightforward extension of the binomial model of Cox, Ross and Rubinstein [2] in which the bond is replaced by a general risky security. It is in this sense that the definitions of the evolution operators $\mathbf{V}_d^B(t,T)$ and $\mathbf{V}_f^B(t,T)$ are consistent with this binomial model. This is the reason that the market environments $\mathbf{V}_d^B$ and $\mathbf{V}_f^B$ with these evolution operators were called in [18] the binomial market environments.

2.3 The Black and Scholes Market Environment

Consider a foreign exchange market in which trading is allowed at any nonnegative time, that is, with the trading time set $\mathcal{T}$ being the set of nonnegative real numbers.

In this section we present the market environments $\mathbf{V}_d^{BS}$ and $\mathbf{V}_f^{BS}$ with evolution operators $\mathbf{V}_d^{BS}(t,T) : \Pi \to \Pi$ and $\mathbf{V}_f^{BS}(t,T) : \Pi \to \Pi$ with t and T in $\mathcal{T}$ and $t \le T$ corresponding to the Black and Scholes model [1]. These market environments were introduced in [18] and, following this article, we call them the *Black and Scholes market environments*.

In more detail, foreign exchange market environments $\mathbf{V}_d^{BS}$ and $\mathbf{V}_f^{BS}$ are called the Black and Scholes market environment if their evolution operators $\mathbf{V}_d^{BS}(t,T) : \Pi \to \Pi$ and $\mathbf{V}_f^{BS}(t,T) : \Pi \to \Pi$, with t and T in $\mathcal{T}$ and $t \le T$ are defined by

(2.20)
$$(\mathbf{V}_d^{BS}(t,T)h_d)(S) =$$
$$\frac{e^{-r_d\tau}}{\sigma\sqrt{2\pi\tau}} \int_0^\infty x^{-1} e^{-(\log x/S - (r_d - r_f - \sigma^2/2)\tau)^2/2\sigma^2\tau} h_d(x)\, dx,$$

and

(2.21)
$$(\mathbf{V}_f^{BS}(t,T)h_f)(S') =$$
$$\frac{e^{-r_f\tau}}{\sigma\sqrt{2\pi\tau}} \int_0^\infty x^{-1} e^{-(\log x/S' - (r_f - r_d - \sigma^2/2)\tau)^2/2\sigma^2\tau} h_f(x)\, dx,$$

where $\sigma > 0$, r_d and r_f are arbitrary real numbers, $\tau = T - t$, and h_d and h_f are admissible payoffs in Π, and where the case of t equal to T is understood as the corresponding limit.

It is clear that the Black and Scholes market environments $\mathbf{V}_d^{BS}$ and $\mathbf{V}_f^{BS}$ are time-homogeneous market environments.

We note that the admissible payoffs in Π can be, for example, characterized by their regularity such as continuity or measurability, and their growth at zero and at infinity such as inverse and direct power growth. Detailed analysis of this admissibility issue based on the semigroup theory is presented in [19].

Financially, r_d and r_f are the continuously compounded interest rates in currency d and currency f, respectively.

If a model of a foreign exchange market is chosen that treats the dynamics of the exchange rates as a stochastic process, then the exchange rates in the market environments $\mathbf{V}_d^{BS}$ and $\mathbf{V}_f^{BS}$ with the evolution operators $\mathbf{V}_d^{BS}(t,T)$ and $\mathbf{V}_f^{BS}(t,T)$ follow a geometric Wiener process with volatility σ. With this financial interpretation, the market environments $\mathbf{V}_d^{BS}$ and $\mathbf{V}_f^{BS}$ correspond to the well known Black and Scholes model [1] in the following sense. The function $(\mathbf{V}_d^{BS}(t,T)h_d)(S)$ of S and t and the function $(\mathbf{V}_f^{BS}(t,T)h_f)(S')$ of S' and t for admissible h_d and h_f in Π are the mild solutions of the Cauchy problem for the Black and Scholes equations

$$\frac{\partial}{\partial t}V_d + \frac{1}{2}\sigma^2 S^2 \frac{\partial^2}{\partial S^2}V_d + (r_d - r_f)S\frac{\partial}{\partial S}V_d - r_d V_d = 0, \quad S > 0,\ t \in [0,T),$$
$$V_d|_{t=T} = h_d,$$

and

$$\frac{\partial}{\partial t}V_f + \frac{1}{2}\sigma^2 S'^2 \frac{\partial^2}{\partial S'^2}V_f + (r_f - r_d)S'\frac{\partial}{\partial S'}V_f - r_f V_f = 0, \quad S' > 0, \ t \in [0,T),$$

$$V_f|_{t=T} = h_f.$$

In this regard the kernels $\mathbf{V}_d^{BS}(t,T)(S_t, S_T)$ and $\mathbf{V}_f^{BS}(t,T)(S'_t, S'_T)$ of the evolution operators $\mathbf{V}_d^{BS}(t,T)$ and $\mathbf{V}_f^{BS}(t,T)$ given by (2.20) and (2.21) are the fundamental solutions of the Black and Scholes equations. These are the reasons that the market environments $\mathbf{V}_d^{BS}$ and $\mathbf{V}_f^{BS}$ with these evolution operators were called in [18] the Black and Scholes market environments.

We comment that while for the unique specification of a solution of a general backward parabolic equation in the upper-half plane it is necessary, in addition to the final condition and certain growth restriction at infinity, to specify the boundary condition at zero, this is not the case for the Black and Scholes equations. For these equations it is enough to specify only the final condition at time T and certain growth restrictions on V_d and V_f for S and S' at zero and at infinity. Roughly speaking, the reason for this is that the coefficients of the partial derivatives of V_d and V_f with respect to the exchange rates vanish in an appropriate manner as the exchange rates approaches zero. Moreover, these growth restrictions could be ensured by imposing on the final payoffs h_d and h_f certain growth conditions for S and S' at zero and at infinity, along with regularity conditions. (See, for example, [35],[4], and [17].)

We note that the Black and Scholes market environment can be defined for the case of an option market with several underlying securities, and where the volatilities, correlation coefficients and the interest rates are general functions of the prices of the underlying securities and time, although the evolution operators, in general, can not be given explicitly (see [18]).

As an illustration we present here the explicit expressions for the values of the foreign exchange European call and put options in the Black and Scholes market environments $\mathbf{V}_d^{BS}$ and $\mathbf{V}_f^{BS}$. These values are given by the celebrated Black and Scholes formulas in [1] extended to the foreign exchange setting by Garman and Kohlhagen [7] and Grabbe [9] both in 1983:

(2.22)
$$C_d^{BS}(t,T,S,X) = C(T-t,S,X,r_d,r_f,\sigma),$$
$$P_d^{BS}(t,T,S,X) = P(T-t,S,X,r_d,r_f,\sigma),$$

and

(2.23)
$$C_f^{BS}(t,T,S',X') = C(T-t,S',X',r_f,r_d,\sigma),$$
$$P_f^{BS}(t,T,S',X') = P(T-t,S',X',r_f,r_d,\sigma),$$

where

$$C(T - t, s, x, \alpha, \beta, \sigma) = e^{-\beta(T-t)} s N(d_+) - e^{-\alpha(T-t)} x N(d_-),$$

$$P(T - t, s, x, \alpha, \beta, \sigma) = e^{-\alpha(T-t)} x N(-d_-) - e^{-\beta(T-t)} s N(-d_+),$$

with

$$d_+ = \frac{\log(s/x) + (\alpha - \beta + \frac{1}{2}\sigma^2)(T - t)}{\sigma\sqrt{T - t}},$$

$$d_- = \frac{\log(s/x) + (\alpha - \beta - \frac{1}{2}\sigma^2)(T - t)}{\sigma\sqrt{T - t}},$$

$$N(x) = \frac{1}{\sqrt{2\pi}} \int_{-\infty}^{x} e^{-y^2/2} \, dy,$$

and the cases of s or x equal to zero are understood as the corresponding limits.

2.4　Relationship Between the Binomial and Black and Scholes Market Environments

It is well known that the values of European options in the Black and Scholes model can be approximated to any level of accuracy by the values of these European options in the binomial model. This was originally shown by Cox, Ross, and Rubenstein in [2]. Here, following [18], we will restate this result as the fact that the Black and Scholes market environment can be viewed as a certain limiting case of a particular sequence of binomial market environments in which the meshes of the trading time sets tend to zero. In this way, we can model the Black and Scholes market environment with a continuous trading time set by a binomial market environment with a discrete trading time set to any desired level of precision.

For fixed times t and T with $0 \leq t < T$, define a sequence of particular binomial market environments $\mathbf{V}_d^{B_n}$ and $\mathbf{V}_f^{B_n}$ with $n = 1, 2, \cdots$, as follows. Each market environment $\mathbf{V}_d^{B_n}$ and $\mathbf{V}_f^{B_n}$ has a trading time set $\mathcal{T}_n = \{t_i = t + i\Delta_n : i = 0, 1, \cdots, n\}$, where $\Delta_n = (T - t)/n$. Note that $t_0 = t$ and $t_n = T$ for each market environment $\mathbf{V}_d^{B_n}$ and $\mathbf{V}_f^{B_n}$.

The market environment $\mathbf{V}_d^{B_n}$ consists of evolution operators $\mathbf{V}_d^{B_n}(t, T)$ defined in (2.5) with the following particular choice of the functions u_l, v_l, $\rho_{u,l}$ and $\rho_{v,l}$:

$$u_l(S) = \exp(\sigma\sqrt{\Delta_n}), \quad v_l(S) = \exp(-\sigma\sqrt{\Delta_n}),$$

$$\rho_{u,l}(S) = \rho_{v,l}(S) = \exp(r_d\Delta_n),$$

where $l = 0, 1, \ldots, n - 1$. Note that these functions do not depend upon the exchange rates and the time index l.

Similarly, the market environment $\mathbf{V}_f^{B_n}$ consists of evolution operators $\mathbf{V}_f^{B_n}(t, T)$ defined in (2.5) with the following particular choice of the functions u_l', v_l', $\rho_{u,l}'$ and $\rho_{v,l}'$:

$$u_l'(S') = \exp(-\sigma\sqrt{\Delta_n}), \quad v_l'(S') = \exp(\sigma\sqrt{\Delta_n}),$$
$$\rho_{u,l}'(S') = \rho_{v,l}'(S') = \exp(r_f\Delta_n),$$

where $l = 0, 1, \ldots, n - 1$. Note that these functions do not depend upon the exchange rates and the time index l.

It is clear that the binomial market environments $\mathbf{V}_d^{B_n}$ and $\mathbf{V}_f^{B_n}$ are time-homogeneous market environments.

As the meshes Δ_n of the trading time sets $\mathcal{T}_n$ tend to zero, or equivalently, as n tends to infinity, the preceding binomial market environments $\mathbf{V}_d^{B_n}$ and $\mathbf{V}_f^{B_n}$ tend to the Black and Scholes market environments $\mathbf{V}_d^{BS}$ and $\mathbf{V}_f^{BS}$ in the following sense

$$\lim_{n\to\infty} \mathbf{V}_d^{B_n}(t, T) = \mathbf{V}_d^{BS}(t, T) \quad \text{and} \quad \lim_{n\to\infty} \mathbf{V}_f^{B_n}(t, T) = \mathbf{V}_f^{BS}(t, T).$$

For the details on these limits see [18]. According to these relationships, operators $\mathbf{V}_d^{BS}(t, T)$ and $\mathbf{V}_f^{BS}(t, T)$ can be approximated to any level of accuracy by the operators $\mathbf{V}_d^{B_n}(t, T)$ and $\mathbf{V}_f^{B_n}(t, T)$. Hence the restrictions $\hat{\mathbf{V}}_d^{B_n}(t, T)$ and $\hat{\mathbf{V}}_f^{B_n}(t, T)$ of the operators $\mathbf{V}_d^{B_n}(t, T)$ and $\mathbf{V}_f^{B_n}(t, T)$ to the binomial trees Γ_d^n and Γ_f^n can be viewed, for sufficiently large n, as approximations to the operators $\mathbf{V}_d^{BS}(t, T)$ and $\mathbf{V}_f^{BS}(t, T)$. In this way, as the meshes Δ_n of the trading time sets $\mathcal{T}_n$ tend to zero, or equivalently, as n tends to infinity, the restricted binomial market environments

$$\hat{\mathbf{V}}_d^{B_n} = \{\hat{\mathbf{V}}_d^{B_n}(t, T) : \Pi_d(T) \to \Pi_d(t) \,|\, t, T \in \mathcal{T}, t \le T\}$$

and

$$\hat{\mathbf{V}}_f^{B_n} = \{\hat{\mathbf{V}}_f^{B_n}(t, T) : \Pi_f(T) \to \Pi_f(t) \,|\, t, T \in \mathcal{T}, t \le T\}$$

tend to the Black and Scholes market environments $\mathbf{V}_d^{BS}$ and $\mathbf{V}_f^{BS}$.

The practical significance of this is that it is the operators $\hat{\mathbf{V}}_d^{B_n}(t, T)$ and $\hat{\mathbf{V}}_f^{B_n}(t, T)$ that are mainly used, implicitly or explicitly, for the numerical evaluation of European foreign exchange options in the Black and Scholes market environment. These approximation properties remain true for the Black and Scholes market environment even if volatility and interest rates are functions of time and exchange rates. In this case, the use of the operators $\hat{\mathbf{V}}_d^{B_n}(t, T)$ and $\hat{\mathbf{V}}_f^{B_n}(t, T)$ for the numerical evaluation of European foreign exchange options becomes explicit.

Equipped with the framework of a market environment we are now in a position to introduce our symmetry in the foreign exchange market.

Chapter 3

Symmetry in a Foreign Exchange Market

The symmetry relationships we wish to derive for the various types of options in a foreign exchange market are based on symmetry relationships between financially equivalent payoffs on the opposite sides of the market. Such financially equivalent payoffs represent the same amount of wealth contingent upon the state of the market as expressed by the values of the exchange rates. This wealth could be thought of as an abstract payoff, which is an element of the space of all abstract payoffs, versus the actual payoffs, which could be received on either side of the foreign exchange market. We formalize the act of observing an abstract payoff via its actual payoff on either the domestic or foreign side of the market as a particular choice of a basis, the domestic or foreign basis, for the space of all abstract payoffs. In the chosen domestic or foreign basis, the coordinates of the abstract payoff, indexed by the corresponding exchange rate, are the values of its domestic or foreign actual payoff for this exchange rate. In this formalism the act of changing sides in a foreign exchange market is described as a change of these bases. The role of the change-of-basis operator is played by the one-dimensional Kelvin transform. It is in terms of this operator that we formulate our symmetry in a foreign exchange market.

For example, the evolution operators determining the market environments in the domestic and foreign sides of the market are, in fact, expressions in the domestic and foreign bases of certain underlying evolution operators and hence related by the similarity transform. These underlying evolution operators are defined on the space of all abstract payoffs and we call them the abstract evolution operators. In this regard, the symmetry relationships for European options are nothing but the similarity relationships for the evolution operators and associate the financially equivalent European options on the opposite sides of the foreign exchange market. These symmetry relationships state that foreign exchange European options on the opposite sides of the market are financially equivalent provided that their payoffs are financially equivalent. In this chapter we also present similar symmetry relationships for foreign exchange Bermudan and American options.

Similarly to the concepts of an abstract payoff and an abstract evolution operator, the symmetry relationships for European, Bermudan, and American options in a foreign exchange market give rise to the concepts of abstract European, Bermudan and American options. These concepts of an abstract payoff, an abstract evolution operator, and abstract European, Bermudan and American options allow for a coordinate-free, that is, (domestic-foreign) basis-independent description of financial phenomena in a foreign exchange market.

In order to maintain clarity from a financial perspective, we present results in this and the next two chapters without their formal mathematical proofs, supporting them whenever possible with appropriate financial justification. Formal mathematical proofs will be given in Part II.

3.1   The Kelvin Transform

We now present the symmetry in a foreign exchange market at the level of payoffs, namely the symmetry that associates the financially equivalent payoffs on the opposite sides of the market. We show that this association is accomplished by the Kelvin transform.

The (one-dimensional) *Kelvin transform* $\mathbf{K}_c : \Pi \to \Pi$ is defined by

$$(\mathbf{K}_c g)(x) = (x/\sqrt{c})g(c/x), \quad x > 0,$$

where $c > 0$ (see Vladimirov, [34]). When $c = 1$, we write $\mathbf{K}_c$ simply as $\mathbf{K}$.

We note that the Kelvin transform preserves the nonnegative cone Π_+ in Π.

Assume that we can receive an amount of wealth in either the domestic side or the foreign side of the foreign exchange market contingent upon the current state of the foreign exchange market as expressed by the values of the exchange rates. This wealth, which could be positive or negative, is independent of the particular currency in which it is measured. If the wealth is to be received in the domestic side of the market, that is, in currency d, contingent upon the exchange rate S between currency d and currency f, then the wealth is a payoff $h_d(S)$ in currency d as a function of the exchange rate S. Similarly, if the wealth is to be received in the foreign side of the market, that is, in currency f, contingent upon the exchange rate S' between currency f and currency d, then the wealth is a payoff $h_f(S')$ in currency f as a function of the exchange rate S'. Because h_d and h_f are real-valued functions on the set of positive real numbers, they are members of the function space Π.

Since the payoffs h_d and h_f express the same amount of wealth, they cannot be independent. They are, in fact, related via the Kelvin transform:

$$(3.1) \qquad\qquad h_d = \mathbf{K}\,h_f, \quad h_f = \mathbf{K}\,h_d.$$

These are the sought after symmetry relationships in a foreign exchange market at the level of payoffs. The proof of these symmetry relationships is based on the following financial argument.

Assume that we receive in the foreign side of the foreign exchange market a payoff $h_f(S')$ in currency f. By converting this payoff to currency d, we receive the payoff $S\,h_f(S')$. By the no-arbitrage argument, $S' = 1/S$ and so we can express the payoff $S\,h_f(S')$ solely in terms of S as $S\,h_f(1/S)$. Since the payoffs h_d and h_f express the same amount of wealth, the payoffs $h_d(S)$ and $S\,h_f(1/S)$ also express the same amounts of wealth, but now both are received on the same side of the foreign exchange market, the domestic side, and both are contingent upon the same exchange rate S. Therefore, because of the no-arbitrage argument, we obtain $h_d(S) = S\,h_f(1/S)$ for all $S > 0$. Since, by the definition of the Kelvin transform, $S\,h_f(1/S) = (\mathbf{K}h_f)(S)$ for all $S > 0$, we obtain the first equality of (3.1).

Similarly, assume that we receive in the domestic side of the foreign exchange market a payoff $h_d(S)$ in currency d. By converting this payoff to currency f, we receive the payoff $S'\,h_d(S)$. By the no-arbitrage argument, $S = 1/S'$ and so we can express the payoff $S'\,h_d(S)$ solely in terms of S' as $S'\,h_d(1/S')$. Since the payoffs h_d and h_f express the same amount of wealth, the payoffs $S'\,h_d(1/S')$ and $h_f(S')$ also express the same amounts of wealth, but now both are received on the same side of the foreign exchange market, the foreign side, and both are contingent upon the same exchange rate S'. Therefore, because of the no-arbitrage argument, we obtain $h_f(S') = S'\,h_d(1/S')$ for all $S' > 0$. Since, by the definition of the Kelvin transform, $S'\,h_d(1/S') = (\mathbf{K}h_d)(S')$ for all $S' > 0$, we obtain the second equality of (3.1).

It is easy to see, that in the particular case of the payoffs of the call and put options the symmetry relationships in (3.1) take the form

$$\mathbf{K}\,(\cdot - X)^+ = X(1/X - \cdot)^+,$$
(3.2)
$$\mathbf{K}\,(X - \cdot)^+ = X(\cdot - 1/X)^+,$$

and

$$\mathbf{K}\,(\cdot - X')^+ = X'(1/X' - \cdot)^+,$$
(3.3)
$$\mathbf{K}\,(X' - \cdot)^+ = X'(\cdot - 1/X')^+,$$

where

$$(\cdot - X)^+ \quad \text{and} \quad (X - \cdot)^+$$

are the call and put payoffs in currency d, and

$$(\cdot - X')^+ \quad \text{and} \quad (X' - \cdot)^+$$

are the call and put payoffs in currency f.

These ideas can be formulated in the following natural abstract framework suggested by modern theoretical physics. Even though the wealth itself can be directly observed only through payoffs h_d and h_f, it is the wealth that has an objective financial status and hence, using the terminology of modern theoretical physics, is a *financial observable*. In this regard, this wealth could be thought of as an abstract payoff h in the space of all abstract payoffs Π in contrast to the payoffs h_d and h_f, which are actual payoffs. It is easy to see that the space of all abstract payoffs Π is a real vector space and, hence, the abstract payoffs themselves, as elements of this space, are vectors. Indeed, the real vector space structure on Π can be established by trivial arguments of the kind that financially it is always possible for any two abstract payoffs to form an abstract payoff which is their sum, and for any abstract payoff and for any real number it is always possible to form an abstract payoff which is their product. The nonnegative cone Π_+ of Π, with the cone defined as for a real vector space, contains precisely those abstract payoffs h that represent an amount of contingent wealth that is nonnegative no matter what is the state of the foreign exchange market. Therefore, it is clear that for any abstract payoff h in Π_+, the actual payoffs h_d and h_f have to be, in turn, members of the nonnegative cone Π_+ of Π. Therefore, for consistency we must require that the Kelvin transform as the operator that maps h_d to h_f, and vice versa, must preserve the nonnegative cone Π_+ of Π. This is indeed the case, as was noted above.

We comment that, as it was shown by Kholodnyi in [18], in a general market with exchange of arbitrary underlying securities the space of all payoffs contingent upon the state of the market as expressed by the prices of these underlying securities, from the financial stand point, must be a real vector space, in fact, a vector lattice. Moreover, it was shown in [18] that the real vector space structure, in fact the vector lattice structure, is necessary for a space of all generalized payoffs, that is, payoffs understood in the generalized sense as a measure of satisfaction, or more precisely, of a marginal utility. Roughly speaking, the arguments in this article for the space of all payoffs were as follows. The real vector space structure can be established by trivial arguments of the kind that financially it is always possible for any two payoffs to form a payoff which is their sum, and for any payoff and for any real number it is always possible to form a payoff which is their product. To establish the vector lattice structure we note that financially it is always possible to distinguish the nonnegative payoffs, that is, such payoffs that have a nonnegative value no matter what is the state of the market. The set of these nonnegative payoffs forms a nonnegative cone in the space of all payoffs. The vector space of all payoffs with the partial order generated by the nonnegative cone is a partially ordered vector space. Then, financially, for any two given payoffs one can always chose a unique payoff that is their supremum, that is, a payoff which is the least among all payoffs greater than or equal to these two payoffs. This means that the lattice operation of supremum must be defined on the space of all payoffs. In turn, in a partially ordered vector space in which a supremum operation is defined, the lattice operation of infimum for any two given payoffs can be defined as the supremum, taken with the opposite sign, of these two payoffs also taken with the opposite signs.

In this way the vector space of all actual payoffs Π with the partial order generated by the nonnegative cone Π_+ is also a partially ordered vector space. Moreover, Π is a vector lattice with the natural operations of supremum and infimum defined as the pointwise maximum and minimum.

In the same way, the vector space of all abstract payoffs $\mathbf{\Pi}$ viewed as a space of all generalized payoffs is also a vector lattice. Indeed, the vector space of all abstract payoffs $\mathbf{\Pi}$ with the partial order generated by the nonnegative cone $\mathbf{\Pi}_+$ is a partially ordered vector space. We define the lattice operation of supremum $\vee$ and infimum $\wedge$ as induced by those in Π. More precisely, for any abstract payoffs x and y the abstract payoff $z = x \vee y$ has its actual payoffs z_d and z_f such that

$$z_d(S) = \max\{x_d(S), y_d(S)\} \quad \text{and} \quad z_f(S') = \max\{x_f(S'), y_f(S')\},$$

for all $S > 0$ and $S' > 0$, where x_d and x_f are actual payoffs of x, and y_d and y_f are actual payoffs of y. Similarly, for any abstract payoffs x and y the abstract payoff $z = x \wedge y$ has its actual payoffs z_d and z_f such that

$$z_d(S) = \min\{x_d(S), y_d(S)\} \quad \text{and} \quad z_f(S') = \min\{x_f(S'), y_f(S')\},$$

for all $S > 0$ and $S' > 0$.

Later in this Chapter we introduce the concepts of European, Bermudan and American options with abstract payoffs which we call abstract European, Bermudan and American options. Abstract European, Bermudan and American options can be viewed and explicitly valued as generalized European, Bermudan and American options, that is, European, Bermudan and American options with generalized payoffs. The generalized European, Bermudan and American options were introduced and explicitly valued in [18]. As was noted in this article, the vector lattice structure on a space of all, possibly generalized, payoffs is necessary for the explicit valuation of Bermudan and American, possibly generalized, options. The explicit valuation of abstract European, Bermudan and American options will be used in the proofs of the symmetry relationships for foreign exchange European, Bermudan and American options in Chapter 8.

It is easy to see that the maps from Π to Π that associate to an abstract payoff h the actual payoffs h_d and h_f are isomorphisms of vector lattices. Therefore, the vector lattice Π of all abstract payoffs is isomorphic to the vector lattice Π.

It is also easy to see that the Kelvin transform $\mathbf{K}$ is an isomorphism of the vector lattice Π.

We formalize the act of observing an abstract payoff via its actual payoff on either side of the foreign exchange market as a particular choice of a basis for the space of all abstract payoffs Π. In the domestic side of the foreign exchange market, this basis, the domestic basis, is chosen in such a way that the coordinates, indexed by the exchange rate S, of the abstract payoff h in Π with respect to this basis are the values of an actual payoff $h_d(S)$. In this way the actual payoff h_d, as a function of the exchange rate S, is the coordinate description of the abstract payoff h in the domestic basis. In the foreign side of the foreign exchange market, this basis, the foreign basis, is chosen in such a way that the coordinates, indexed by the exchange rate S', of the abstract payoff h with respect to this basis are the values of an actual payoff $h_f(S')$. In this way the actual payoff h_f, as a function of the exchange rate S', is the coordinate description of the abstract payoff h in the foreign basis.

In this formalism, the act of changing sides in a foreign exchange market is described by a linear change-of-basis operator that acts on the space Π or, more precisely, by an isomorphism of Π as a vector lattice. Due to the non-preferability of each side of the foreign exchange market, this operator has to perform both the change of basis from domestic to foreign and the change of basis from foreign to domestic. In addition, by performing the change of basis twice, starting with either side of the market, we have to restore the original basis. Therefore, the change-of-basis operator has to be a faithful representation of the cyclic group $\mathbb{Z}_2$ on Π.

In order to find the explicit form of the change-of-basis operator we note that the definition of a change-of-basis operator implies that, for any abstract payoff, the operator has to map the expression of this payoff in the domestic basis, that is, the domestic actual payoff, to the expression of this abstract payoff in the foreign basis, that is, to the foreign actual payoff, and vice versa. This means that for each abstract payoff h in Π, the change-of-basis operator has to map the domestic actual payoff h_d in Π to the foreign actual payoff h_f in Π, and the foreign actual payoff h_f in Π to the domestic actual payoff h_d in Π. Thus the equalities (3.1) imply that it is the Kelvin transform that is the change-of-basis operator. As it is easy to verify, the Kelvin transform $\mathbf{K}$ as a linear operator on Π or, more precisely, an isomorphism of Π as a vector lattice, is indeed a faithful representation of the cyclic group $\mathbb{Z}_2$ on Π.

The $\mathbb{Z}_2$ symmetry in the foreign exchange market at the level of payoffs has its counterpart at the level of foreign exchange options. We start with European options.

3.2   Symmetry for European Options

In this section we extend the $\mathbb{Z}_2$ symmetry in the foreign exchange market at the level of payoffs to the symmetry at the level of foreign exchange European options. In order to do this we place the concept of a European option into the abstract setting introduced in the previous section.

An *abstract European option* with inception time t, expiration time T (t and T are in $\mathcal{T}$, and $t \leq T$), and with an abstract payoff g in Π_+ is a contract that gives the right, but not the obligation, to receive the abstract payoff g at the expiration time T.

For each t and T in $\mathcal{T}$ with $t \leq T$, denote by $\mathcal{V}(t,T)$ the operator that maps an abstract payoff g of an abstract European option with inception time t and expiration time T to the abstract value of the option denoted by $\mathcal{E}(t,T,g)$ at the inception time t, that is

$$\mathcal{E}(t,T,g) = \mathcal{V}(t,T)\,g,$$

where g is in Π_+. Since, by the no-arbitrage argument, the value of an abstract European option at its expiration time is equal to its abstract payoff, $\mathcal{V}(t,T)$

is the identity operator whenever inception time t is equal to expiration time T.

By definition, the domain of the operator $\mathcal{V}(t,T)$ is in $\mathbf{\Pi}_+$. Due to the no-free-lunch argument, the range of the operator $\mathcal{V}(t,T)$ is also in $\mathbf{\Pi}_+$. Moreover, the operator $\mathcal{V}(t,T) : \mathbf{\Pi}_+ \to \mathbf{\Pi}_+$ admits a unique extension to an operator whose domain is a linear subspace in $\mathbf{\Pi}$ and whose range is in $\mathbf{\Pi}$. Denote this new operator again by $\mathcal{V}(t,T)$. We call $\mathcal{V}(t,T)$ an *abstract evolution operator*. Due to the no-arbitrage argument, the operator $\mathcal{V}(t,T) : \mathbf{\Pi} \to \mathbf{\Pi}$ is a linear operator and, due to the previous discussion, it preserves the nonnegative cone $\mathbf{\Pi}_+$ in $\mathbf{\Pi}$. Moreover, due to the previous paragraph, $\mathcal{V}(t,t)$ is the identity operator on $\mathbf{\Pi}$ for each t in $\mathcal{T}$.

We call the family $\mathcal{V}$ of abstract evolution operators

$$\mathcal{V} = \{\mathcal{V}(t,T) : \mathbf{\Pi} \to \mathbf{\Pi} \,|\, t, T, \in \mathcal{T}, t \leq T\}$$

an abstract market environment.

In the domestic basis for $\mathbf{\Pi}$ the abstract evolution operator $\mathcal{V}(t,T)$ is given by $\mathbf{V}_d(t,T)$ and in the foreign basis for $\mathbf{\Pi}$ the abstract evolution operator is given by $\mathbf{V}_f(t,T)$. Since $\mathbf{V}_d(t,T)$ and $\mathbf{V}_f(t,T)$ are the same operator $\mathcal{V}(t,T)$ expressed in different bases with the change-of-basis operator being the Kelvin transform $\mathbf{K}$, they must be related by the similarity transform:

$$\text{(3.4)} \qquad \begin{aligned} \mathbf{V}_d(t,T) &= \mathbf{K}\,\mathbf{V}_f(t,T)\,\mathbf{K}, \\ \mathbf{V}_f(t,T) &= \mathbf{K}\,\mathbf{V}_d(t,T)\,\mathbf{K}, \end{aligned}$$

where we have made use of the fact that $\mathbf{K}^{-1} = \mathbf{K}$, since $\mathbf{K}$ is a faithful representation of the group $\mathbb{Z}_2$ on Π.

These relationships have the financial interpretation of a symmetry that associates financially equivalent European options on the opposite sides of the foreign exchange market, and state that foreign exchange European options on the opposite sides of the market are financially equivalent provided that their payoffs are financially equivalent.

In more detail, the expression of $\mathcal{E}(t,T,g)$, as an element of $\mathbf{\Pi}$, even of $\mathbf{\Pi}_+$, in the domestic basis is the value $\mathcal{E}_d(t,T,g_d)$ in Π_+ of the European option in currency d on currency f with inception time t, expiration time T and with payoff g_d, where g_d is the expression of the abstract payoff g in the domestic basis. On the other hand, the expression of $\mathcal{E}(t,T,g)$ in the foreign basis is the value $\mathcal{E}_f(t,T,g_f)$ in Π_+ of the European option in currency f on currency d with inception time t, expiration time T, and with payoff g_f, where g_f is the expression of the abstract payoff g in the foreign basis. Therefore, for any g_d

and g_f in Π_+ the relationships (3.4) read

$$
(3.5) \quad
\begin{aligned}
\mathcal{E}_d(t,T,g_d) &= \mathbf{K}\,\mathcal{E}_f(t,T,\mathbf{K}\,g_d), \quad g_d \in \Pi_+, \\
\mathcal{E}_f(t,T,g_f) &= \mathbf{K}\,\mathcal{E}_d(t,T,\mathbf{K}\,g_f), \quad g_f \in \Pi_+.
\end{aligned}
$$

In words, the value $\mathcal{E}_f(t,T,g_f)$, converted to currency d, of the European option in currency f on currency d with payoff g_f has to be equal to the value $\mathcal{E}_d(t,T,g_d)$ of the European option in currency d on currency f with payoff g_d, where g_f is the payoff g_d converted to currency f. Otherwise, there must be an arbitrage opportunity in the foreign exchange market. Similarly, the value $\mathcal{E}_d(t,T,g_d)$, converted to currency f, of the European option in currency d on currency f with payoff g_d has to be equal to the value $\mathcal{E}_f(t,T,g_f)$ of the European option in currency f on currency d with payoff g_f, where g_d is the payoff g_f converted to currency d.

In the particular case of European call and put options, the symmetry relationships (3.5) take the form of the relationships (1.1) and (1.2).

For any given model of foreign exchange option markets, the symmetry relationships (3.4) present an additional requirement, a consistency requirement, and therefore they should be verified directly on a case-by-case basis. For example, as will be shown in Chapter 7, in the two major cases, the binomial and the Black and Scholes market environments, this consistency requirement is satisfied.

More precisely, the evolution operators $\mathbf{V}_d^B(t,T)$ and $\mathbf{V}_f^B(t,T)$ for the binomial market environment, and their restrictions $\hat{\mathbf{V}}_d^B(t,T)$ and $\hat{\mathbf{V}}_f^B(t,T)$ to the binomial trees Γ_d^n and Γ_f^n satisfy the relationships (3.4), that is,

$$
(3.6) \quad
\begin{aligned}
\mathbf{V}_d^B(t,T) &= \mathbf{K}\,\mathbf{V}_f^B(t,T)\,\mathbf{K}, \\
\mathbf{V}_f^B(t,T) &= \mathbf{K}\,\mathbf{V}_d^B(t,T)\,\mathbf{K},
\end{aligned}
$$

and

$$
(3.7) \quad
\begin{aligned}
\hat{\mathbf{V}}_d^B(t,T) &= \hat{\mathbf{K}}_{f\to d}(t)\,\hat{\mathbf{V}}_f^B(t,T)\,\hat{\mathbf{K}}_{d\to f}(T), \\
\hat{\mathbf{V}}_f^B(t,T) &= \hat{\mathbf{K}}_{d\to f}(t)\,\hat{\mathbf{V}}_d^B(t,T)\,\hat{\mathbf{K}}_{f\to d}(T),
\end{aligned}
$$

where $\hat{\mathbf{K}}_{d\to f}(t) : \Pi_d(t) \to \Pi_f(t)$ and $\hat{\mathbf{K}}_{f\to d}(t) : \Pi_f(t) \to \Pi_d(t)$ are the restrictions of the Kelvin transform $\mathbf{K}$ defined by

$$
(3.8) \quad Q_d(t)\mathbf{K} = \hat{\mathbf{K}}_{f\to d}(t)Q_f(t), \quad Q_f(t)\mathbf{K} = \hat{\mathbf{K}}_{d\to f}(t)Q_d(t).
$$

Note that

$$
(3.9) \quad \hat{\mathbf{K}}_{f\to d}(t)\hat{\mathbf{K}}_{d\to f}(t) = \mathbf{I}_d(t), \quad \hat{\mathbf{K}}_{d\to f}(t)\hat{\mathbf{K}}_{f\to d}(t) = \mathbf{I}_f(t),
$$

where $\mathbf{I}_d(t)$ and $\mathbf{I}_f(t)$ are the identity operators on $\Pi_d(t)$ and $\Pi_f(t)$ respectively.

The evolution operators $\mathbf{V}_d^{BS}(t,T)$ and $\mathbf{V}_f^{BS}(t,T)$ for the Black and Scholes market environment also satisfy the symmetry relationships (3.4), that is,

(3.10)
$$\mathbf{V}_d^{BS}(t,T) = \mathbf{K}\,\mathbf{V}_f^{BS}(t,T)\,\mathbf{K},$$
$$\mathbf{V}_f^{BS}(t,T) = \mathbf{K}\,\mathbf{V}_d^{BS}(t,T)\,\mathbf{K}.$$

Although the symmetry relationships (1.1) and (1.2) for European call and put options in the Black and Scholes market environment are a particular case of the symmetry relationships (3.5) and symmetry relationships (3.4) as given by (3.10) and as such will be proved in Part II, we derive them now, as an illustration, directly from the well-known explicit expressions (2.22) and (2.23) for these call and put options given by the Black and Scholes formulas. More precisely, we show that

$$C_d^{BS}(t,T,S,X) = SX\,P_f^{BS}(t,T,1/S,1/X),$$
$$P_d^{BS}(t,T,S,X) = SX\,C_f^{BS}(t,T,1/S,1/X),$$

$$C_f^{BS}(t,T,S',X') = S'X'\,P_d^{BS}(t,T,1/S',1/X'),$$
$$P_f^{BS}(t,T,S',X') = S'X'\,C_d^{BS}(t,T,1/S',1/X').$$

We prove only the first of the preceding symmetry relationships, since the proof of the remaining relationships is completely analogous. The proof follows from the following chain of equalities:

$$C_d^{BS}(t,T,S,X) = e^{-r_f(T-t)}\,S\,N\left(\frac{\log(S/X) + (r_d - r_f + \frac{1}{2}\sigma^2)(T-t)}{\sigma\sqrt{T-t}}\right)$$

$$- e^{-r_d(T-t)}\,X\,N\left(\frac{\log(S/X) + (r_d - r_f - \frac{1}{2}\sigma^2)(T-t)}{\sigma\sqrt{T-t}}\right)$$

$$= SX\left(e^{-r_f(T-t)}\,X^{-1}\,N\left(\frac{\log(X^{-1}/S^{-1}) + (r_d - r_f + \frac{1}{2}\sigma^2)(T-t)}{\sigma\sqrt{T-t}}\right)\right.$$

$$\left. - e^{-r_d(T-t)}\,S^{-1}\,N\left(\frac{\log(X^{-1}/S^{-1}) + (r_d - r_f - \frac{1}{2}\sigma^2)(T-t)}{\sigma\sqrt{T-t}}\right)\right)$$

$$= SX\left(e^{-r_f(T-t)}\,X^{-1}\,N\left(-\frac{\log(S^{-1}/X^{-1}) + (r_f - r_d - \frac{1}{2}\sigma^2)(T-t)}{\sigma\sqrt{T-t}}\right)\right.$$

$$\left. - e^{-r_d(T-t)}\,S^{-1}\,N\left(-\frac{\log(S^{-1}/X^{-1}) + (r_f - r_d + \frac{1}{2}\sigma^2)(T-t)}{\sigma\sqrt{T-t}}\right)\right)$$

$$= SX\,P_f^{BS}(t,T,1/S,1/X),$$

where the first and the last equalities are due to the expressions (2.22) and (2.23), the second and third equalities are due to trivial algebra.

To present the practical significance of the symmetry relationships (3.7), we rewrite it in terms of values of foreign exchange European options in the binomial market environments $\mathbf{V}_d^B$ and $\mathbf{V}_f^B$ restricted to the binomial trees Γ_d and Γ_f, that is, in the restricted binomial market environments $\hat{\mathbf{V}}_d^B$ and $\hat{\mathbf{V}}_f^B$

(3.11)
$$\hat{\mathcal{E}}_d(t,T,\hat{g}_d) = \hat{\mathbf{K}}_{f\to d}(t)\,\hat{\mathcal{E}}_f(t,T,\hat{\mathbf{K}}_{d\to f}(T)\,\hat{g}_d), \ \ \hat{g}_d \in \Pi_{d_+}(T),$$
$$\hat{\mathcal{E}}_f(t,T,\hat{g}_f) = \hat{\mathbf{K}}_{d\to f}(t)\,\hat{\mathcal{E}}_d(t,T,\hat{\mathbf{K}}_{f\to d}(T)\,\hat{g}_f), \ \ \hat{g}_f \in \Pi_{f_+}(T),$$

where $\Pi_{d_+}(t)$ and $\Pi_{f_+}(t)$ denote the nonnegative cones of $\Pi_d(t)$ and $\Pi_f(t)$, that is, the sets of all nonnegative real-valued functions on $\Gamma_d^n(t)$ and $\Gamma_f^n(t)$, and where

$$\hat{\mathcal{E}}_d(t,T,\hat{g}_d) = \hat{\mathbf{V}}_d^B(t,T)\,\hat{g}_d,$$
$$\hat{\mathcal{E}}_f(t,T,\hat{g}_f) = \hat{\mathbf{V}}_f^B(t,T)\,\hat{g}_f,$$

with the operators $\hat{\mathbf{V}}_d^B(t,T)$ and $\hat{\mathbf{V}}_f^B(t,T)$ given by (2.19).

In the particular case of European call and put options valued in the binomial market environments $\mathbf{V}_d^B$ and $\mathbf{V}_f^B$ restricted to the binomial trees Γ_d and Γ_f, that is, in the restricted binomial market environments $\hat{\mathbf{V}}_d^B$ and $\hat{\mathbf{V}}_f^B$, the symmetry relationships in (3.11) take the form:

$$\hat{C}_d(t,T,S,X) = SX\,\hat{P}_f(t,T,1/S,1/X),$$
$$\hat{P}_d(t,T,S,X) = SX\,\hat{C}_f(t,T,1/S,1/X),$$

and

$$\hat{C}_f(t,T,S',X') = S'X'\,\hat{P}_d(t,T,1/S',1/X'),$$
$$\hat{P}_f(t,T,S',X') = S'X'\,\hat{C}_d(t,T,1/S',1/X'),$$

where S is in $\Gamma_d^n(t)$ and S' is in $\Gamma_f^n(t)$.

As was indicated in Section 2.4, the operators $\hat{\mathbf{V}}_d^{B_n}(t,T)$ and $\hat{\mathbf{V}}_f^{B_n}(t,T)$ can be viewed as approximations of the operators $\mathbf{V}_d^{BS}(t,T)$ and $\mathbf{V}_f^{BS}(t,T)$, and it is the operators $\hat{\mathbf{V}}_d^{B_n}(t,T)$ and $\hat{\mathbf{V}}_f^{B_n}(t,T)$ that are used for the numerical evaluation of European foreign exchange options in the Black and Scholes market environment. Therefore (3.11) indicates that the symmetry relationships hold at all levels of accuracy in numerical evaluation of European foreign exchange options in the Black and Scholes market environment via such approximations.

Now we turn to the symmetry relationships for Bermudan and American options.

3.3   The Intervention Condition

The symmetry relationships for Bermudan and American options appearing later in this Chapter require, for their rigorous proofs, the market environments $\mathbf{V}_d$ and $\mathbf{V}_f$ defined in Section 2.1 to satisfy the intervention condition, introduced by Kholodnyi in [18] and presented below, although in this chapter we rely on financial justification without this assumption.

A market environment $\mathbf{V}$, that is, the family of evolution operators

$$\mathbf{V} = \{\mathbf{V}(t,T) : \Pi \to \Pi \,|\, t, T \in \mathcal{T}, t \leq T\},$$

is said to satisfy the *intervention condition* if for each t, τ, and T in $\mathcal{T}$ such that $t \leq \tau \leq T$ the following relation holds:

$$\mathbf{V}(t,\tau)\mathbf{V}(\tau,T) = \mathbf{V}(t,T).$$

Similarly, the abstract market environment $\mathcal{V}$, that is, the family of abstract evolution operators

$$\mathcal{V} = \{\mathcal{V}(t,T) : \mathbf{\Pi} \to \mathbf{\Pi} \,|\, t, T, \in \mathcal{T}, t \leq T\}$$

is said to satisfy the *abstract intervention condition* if for each t, τ, and T in $\mathcal{T}$ such that $t \leq \tau \leq T$ the following relation holds:

(3.12) $$\mathcal{V}(t,\tau)\mathcal{V}(\tau,T) = \mathcal{V}(t,T).$$

The fact that the abstract market environment $\mathcal{V}$ satisfies the abstract intervention condition means that the market environments $\mathbf{V}_d$ and $\mathbf{V}_f$ also satisfy the intervention condition, that is, for each t, τ, and T in $\mathcal{T}$ such that $t \leq \tau \leq T$ the following relations hold:

$$\mathbf{V}_d(t,\tau)\mathbf{V}_d(\tau,T) = \mathbf{V}_d(t,T),$$

and

$$\mathbf{V}_f(t,\tau)\mathbf{V}_f(\tau,T) = \mathbf{V}_f(t,T).$$

In the relevant foreign exchange setting, the intervention condition financially expresses the requirement of intertemporal no-arbitrage in a foreign exchange option market and is a generalization to a general market environment of a semigroup intertemporal no-arbitrage condition introduced by Garman in [6] which, in our terminology, is applicable only to a time-homogeneous market environment.

If a model of the foreign exchange market is chosen that treats the dynamics of the exchange rates as a stochastic process, then the assumption that the market environments $\mathbf{V}_d$ and $\mathbf{V}_f$ satisfy the intervention condition means that this stochastic process has a Markovian property. For example, the intervention condition holds for the family of evolution operators in the binomial and the Black and Scholes market environments.

A market environment $\mathbf{V}$ that satisfies the intervention condition and such that its trading time set $\mathcal{T}$ is an interval, either finite or infinite, of positive real numbers and its evolution operators $\mathbf{V}(t, T)$ are sufficiently smooth functions of time, admits the following simple characterization given by Kholodnyi in [18], [17] and [20].

We say that that the one-parameter family of linear operators

$$L = \{L(t) : \Pi \to \Pi \,|\, t \in \mathcal{T}\},$$

generates the market environment $\mathbf{V}$ if for each t and T in the trading time set $\mathcal{T}$ with $t \leq T$ and for each admissible payoff v_T in Π the function $\mathbf{V}(t, T)v_T$ of t is a solution, possibly generalized, of the Cauchy problem for the evolution equation

$$\frac{d}{dt}v + L(t)v = 0, \quad t < T,$$
$$v(T) = v_T.$$

(3.13)

An operator $L(t)$ in the family L is called a *generator at time t*, or simply a *generator*.

We comment that the class of admissible payoffs can be characterized by the properties of the evolution operator $\mathbf{V}(t, T)$ as an operator-valued function of t and T and is beyond the scope of this book. The analysis of this admissibility issue is presented in [18].

We also comment that because of the definition of the evolution operators $\mathbf{V}(t, T)$, for each admissible v_T in Π_+ the solution $\mathbf{V}(t, T)v_T$ in Π_+ at time t of the Cauchy problem for the evolution equation in (3.13) is the value at the inception time t of the European option with expiration time T and the payoff v_T in the market environment $\mathbf{V}$. Hence the evolution equation in (3.13) determines the values of European options in this market environment. Approximating the time derivative in this evolution equation by an appropriate finite difference, it is easy to see that the generator $L(t)$ at time t determines an infinitesimal change in the values of such European options at this time t.

Similarly, we say that that the one-parameter family of linear operators

$$\mathcal{L} = \{\mathcal{L}(t) : \Pi \to \Pi \,|\, t \in \mathcal{T}\},$$

generates the abstract market environment $\mathcal{V}$ if for each t and T in the trading time set $\mathcal{T}$ with $t \leq T$ and for each admissible abstract payoff g in Π the function $\mathcal{V}(t, T)g$ of t is a solution, possibly generalized, of the Cauchy problem for the evolution equation

(3.14)
$$\frac{d}{dt}v + \mathcal{L}(t)v = 0, \quad t < T,$$
$$v(T) = g,$$

where we again assume that the trading time set $\mathcal{T}$ is an interval, either finite or infinite, of positive real numbers. We call an operator $\mathcal{L}(t)$ in the family $\mathcal{L}$ an *abstract generator at time t*, or simply an *abstract generator*.

We comment that because of the definition of the abstract evolution operators $\mathcal{V}(t, T)$, for each admissible g in Π_+ the solution $\mathcal{V}(t, T)g$ at time t of the Cauchy problem for the evolution equation in (3.14) is the value at the inception time t of the abstract European option with expiration time T and the abstract payoff g in the abstract market environment $\mathcal{V}$. Hence the evolution equation in (3.14) determines the values of abstract European options in this abstract market environment. Approximating the time derivative in this evolution equation by an appropriate finite difference, it is easy to see that the abstract generator $\mathcal{L}(t)$ at time t determines an infinitesimal change in the values of such abstract European options at this time t.

The fact that an abstract market environment $\mathcal{V}$ is generated by the family $\mathcal{L}$ means that that the market environments $\mathbf{V}_d$ and $\mathbf{V}_f$ are generated by the families of linear operators

$$L_d = \{L_d(t) : \Pi \to \Pi \,|\, t \in \mathcal{T}\},$$

and

$$L_f = \{L_f(t) : \Pi \to \Pi \,|\, t \in \mathcal{T}\},$$

where for each t in the trading time set $\mathcal{T}$ the generator $L_d(t)$ at time t and the generator $L_f(t)$ at time t are the expressions of the abstract generator $\mathcal{L}(t)$

at time t in the domestic and foreign bases for the space of all abstract payoffs Π.

More precisely, for each t and T in the trading time set $\mathcal{T}$ with $t \leq T$ and for each admissible g_d and g_f in Π the functions $\mathbf{V}_d(t,T)g_d$ and $\mathbf{V}_f(t,T)g_f$ of t are solutions, possibly generalized, of the Cauchy problem for the evolution equations

$$
(3.15) \qquad \frac{d}{dt}v_d + \boldsymbol{L}_d(t)v_d = 0, \quad t < T,
$$
$$
v_d(T) = g_d,
$$

and

$$
(3.16) \qquad \frac{d}{dt}v_f + \boldsymbol{L}_f(t)v_f = 0, \quad t < T,
$$
$$
v_f(T) = g_f.
$$

We comment again that because of the definition of the evolution operators $\mathbf{V}_d(t,T)$ and $\mathbf{V}_f(t,T)$, for each admissible g_d and g_f in Π_+, the solutions $\mathbf{V}_d(t,T)g_d$ and $\mathbf{V}_f(t,T)g_f$ at time t of the Cauchy problem for the evolution equations in (3.15) and (3.16) are the values at the inception time t of the European options in currency d on currency f and in currency f on currency d with expiration time T and the payoffs g_d and g_f in the market environments $\mathbf{V}_d$ and $\mathbf{V}_f$. Hence the evolution equations in (3.15) and (3.16) determine the values of European options in these market environments.

For example, in the Black and Scholes market environments $\mathbf{V}_d^{BS}$ and $\mathbf{V}_f^{BS}$ the evolution equations in (3.15) and (3.16) take the form of the Black and Scholes equations presented in Section 2.3, where the generators $\boldsymbol{L}_d^{BS}$ and $\boldsymbol{L}_f^{BS}$ do not depend on time and are given by

$$
\boldsymbol{L}_d^{BS} = \frac{1}{2}\sigma^2 S^2 \frac{\partial^2}{\partial S^2} + (r_d - r_f)S\frac{\partial}{\partial S} - r_d,
$$

and

$$
\boldsymbol{L}_f^{BS} = \frac{1}{2}\sigma^2 S'^2 \frac{\partial^2}{\partial S'^2} + (r_f - r_d)S'\frac{\partial}{\partial S'} - r_f.
$$

We comment that the symmetry relationships (3.4) for the evolution operators $\mathbf{V}_d(t,T)$ and $\mathbf{V}_f(t,T)$ in the market environments $\mathbf{V}_d$ and $\mathbf{V}_f$ under consideration can be reformulated as the symmetry relationships for the generators $\boldsymbol{L}_d(t)$ and $\boldsymbol{L}_f(t)$ of these market environments.

Since for each t in trading time set $\mathcal{T}$ the operators $\boldsymbol{L}_d(t)$ and $\boldsymbol{L}_f(t)$ are the same operator $\mathcal{L}(t)$ expressed in different bases for Π with the change-of-basis operator being the Kelvin transform $\mathbf{K}$, they must be related by the similarity transform

(3.17)
$$L_d(t) = \mathbf{K} \, L_f(t) \, \mathbf{K},$$
$$L_f(t) = \mathbf{K} \, L_d(t) \, \mathbf{K},$$

where we have made use of the fact that $\mathbf{K}^{-1} = \mathbf{K}$, since $\mathbf{K}$ is a faithful representation of the group $\mathbb{Z}_2$ on Π.

For example, the symmetry relationships (3.17) for the case of the Black and Scholes market environments $\mathbf{V}_d^{BS}$ and $\mathbf{V}_f^{BS}$ take the form

(3.18)
$$L_d^{BS} = \mathbf{K} \, L_f^{BS} \, \mathbf{K},$$
$$L_f^{BS} = \mathbf{K} \, L_d^{BS} \, \mathbf{K},$$

and are the reformulation of the symmetry relationships (3.10) for the evolution operators $\mathbf{V}_d^{BS}(t, T)$ and $\mathbf{V}_f^{BS}(t, T)$.

A rigorous proof of the symmetry relationships in (3.17) is based on the analysis of the evolution operators $\mathbf{V}_d(t, T)$ and $\mathbf{V}_f(t, T)$ as operator-valued functions of t and T and is beyond the scope of this book. However, the proof of these symmetry relationships for the case of the Black and Scholes market environments $\mathbf{V}_d^{BS}$ and $\mathbf{V}_f^{BS}$ as given by the symmetry relationships in (3.18) will be presented in Chapter 7.

3.4   Symmetry for Bermudan Options

Now we present the symmetry in the foreign exchange market for Bermudan options with general time-dependent payoffs.

Assume that for t and T in the trading time set $\mathcal{T}$ with $t \leq T$, the *exercise time set* $E = \{t_i : i = 0, 1, \ldots, n\}$ with $t \leq t_0 < t_1 < \ldots < t_{n-1} < t_n = T$ is contained in $\mathcal{T}$.

A *Bermudan option* (on a general underlying security) with inception time t, expiration time T, exercise time set E, and with a (time-dependent) payoff $g : E \to \Pi_+$ is a contract that gives the right, but not the obligation, to receive the payoff $g_\tau(S_\tau)$ at any time τ in the exercise time set E, where the price of the underlying security at time τ is S_τ.

Among Bermudan options, one of the most widely traded are Bermudan call and put options.

A *Bermudan call option* and a *Bermudan put option* are Bermudan options with time-independent payoffs $(\cdot - X)^+ : E \to \Pi_+$ and $(X - \cdot)^+ : E \to \Pi_+$, where $X > 0$ is called the strike price. Financially, Bermudan call and put options give the right but not the obligation to by or to sell one unit of an underlying security at the strike price at any time in the exercise time set.

The payoffs $(\cdot - X)^+ : E \to \Pi_+$ and $(X - \cdot)^+ : E \to \Pi_+$ are called a *call payoff* and a *put payoff* respectively.

We comment that although the call and put payoffs defined for European and for Bermudan options are mathematically different functions, financially they have the same meaning. That is why, whenever ambiguity is not likely, we use the same notations for these call and put payoffs.

In the foreign exchange market in which the underlying security is the domestic or foreign currency, we denote the value of the Bermudan option in currency d on currency f with inception time t, expiration time T, and with payoff g_d in currency d by $\mathcal{B}_d(t, T, E, g_d) = \mathcal{B}_d(t, T, E, g_d)(S_t)$. Similarly, we denote the value of the Bermudan option in currency f on currency d with inception time t, expiration time T, and with payoff g_f in currency f by $\mathcal{B}_f(t, T, E, g_f) = \mathcal{B}_f(t, T, E, g_f)(S'_t)$. Clearly $\mathcal{B}_d(t, T, E, g_d)$ and $\mathcal{B}_f(t, T, E, g_f)$ are in Π_+.

In this notation, the values of the foreign exchange Bermudan call and put options can be represented as

$$C_d(t, T, S, X) = \mathcal{B}_d(t, T, E, (\cdot - X)^+)(S),$$
$$P_d(t, T, S, X) = \mathcal{B}_d(t, T, E, (X - \cdot)^+)(S),$$

and

$$C_f(t, T, S', X') = \mathcal{B}_f(t, T, E, (\cdot - X')^+)(S'),$$
$$P_f(t, T, S', X') = \mathcal{B}_f(t, T, E, (X' - \cdot)^+)(S').$$

As for the preceding case of European options, it is possible to give a coordinate-free, that is, a (domestic-foreign) basis-independent, description of Bermudan options in the foreign exchange market in terms of an abstract Bermudan option.

An *abstract Bermudan option* with inception time t, expiration time T, exercise time set E and with a (time-dependent) abstract payoff $g : E \to \mathbf{\Pi}_+$ is a contract that gives the right, but not the obligation, to receive the abstract payoff g_τ in $\mathbf{\Pi}_+$ at any time τ in the exercise time set E.

Clearly the value $\mathcal{B}(t, T, E, g)$ of this abstract Bermudan option is an element of $\mathbf{\Pi}$, even of $\mathbf{\Pi}_+$. Therefore, the expression of $\mathcal{B}(t, T, E, g)$, as an element of $\mathbf{\Pi}$, in the domestic basis is the value $\mathcal{B}_d(t, T, E, g_d)$ in Π_+ of the Bermudan option in currency d on currency f with inception time t, expiration time T, and with payoff g_d, where g_d is the expression of the abstract payoff g in the domestic basis. On the other hand, the expression of $\mathcal{B}(t, T, E, g)$, as an element of Π, in the foreign basis is the value $\mathcal{B}_f(t, T, E, g_f)$ in Π_+ of the Bermudan option in currency f on currency d with inception time t, expiration time T, and with payoff g_f, where g_f is the expression of the abstract payoff g in the foreign basis.

This is the symmetry we are seeking for foreign exchange Bermudan options. This symmetry associates financially equivalent Bermudan options on the opposite sides of the foreign exchange market, and states that foreign exchange Bermudan options on the opposite sides of the market are financially equivalent provided that their payoffs are financially equivalent.

In more details, the value $\mathcal{B}_f(t, T, E, g_f)$, converted into currency d, of the Bermudan option in currency f on currency d with payoff g_f has to be equal to the value $\mathcal{B}_d(t, T, E, g_d)$ of the Bermudan option in currency d on currency f with payoff g_d, where g_f is the payoff g_d converted into currency f. Otherwise there must be an arbitrage opportunity in the foreign exchange market. Similarly, the value $\mathcal{B}_d(t, T, E, g_d)$, converted into currency f, of the Bermudan option in currency d on currency f with payoff g_d has to be equal to the value $\mathcal{B}_f(t, T, E, g_f)$ of the Bermudan option in currency f on currency d with payoff g_f, where g_d is the payoff g_f converted into currency d. In symbols these symmetry relationships for foreign exchange Bermudan options can be stated as

$$(3.19) \qquad \begin{aligned} \mathcal{B}_d(t, T, E, g_d) &= \mathbf{K}\mathcal{B}_f(t, T, E, \mathbf{K}g_d), \quad g_d : E \to \Pi_+, \\ \mathcal{B}_f(t, T, E, g_f) &= \mathbf{K}\mathcal{B}_d(t, T, E, \mathbf{K}g_f), \quad g_f : E \to \Pi_+. \end{aligned}$$

In the particular case of Bermudan call and put options, the symmetry relationships (3.19) take the form of the relationships (1.1) and (1.2).

The symmetry relationships (3.19) hold for foreign exchange Bermudan options valued in the two major cases of market environments, the binomial market environments $\mathbf{V}_d^B$ and $\mathbf{V}_f^B$ and the Black and Scholes market environments $\mathbf{V}_d^{BS}$ and $\mathbf{V}_f^{BS}$. The symmetry relationships (3.19) also hold for foreign exchange Bermudan options valued in the binomial market environments $\mathbf{V}_d^B$ and $\mathbf{V}_f^B$ restricted to the binomial trees Γ_d^n and Γ_f^n, that is, in the restricted binomial market environments $\hat{\mathbf{V}}_d^B$ and $\hat{\mathbf{V}}_f^B$:

$$(3.20) \qquad \begin{aligned} \hat{\mathcal{B}}_d(t, T, E, \hat{g}_d) &= \hat{\mathbf{K}}_{f \to d}(t)\hat{\mathcal{B}}_f(t, T, E, \hat{\mathbf{K}}_{d \to f}(\cdot)\hat{g}_d), \\ \hat{\mathcal{B}}_f(t, T, E, \hat{g}_f) &= \hat{\mathbf{K}}_{d \to f}(t)\hat{\mathcal{B}}_d(t, T, E, \hat{\mathbf{K}}_{f \to d}(\cdot)\hat{g}_f), \end{aligned}$$

where $\hat{g}_{d,\tau}$ is in $\Pi_{d_+}(\tau)$ and $\hat{g}_{f,\tau}$ is in $\Pi_{f_+}(\tau)$, and $(\hat{\mathbf{K}}_{d \to f}(\cdot)\hat{g}_d)_\tau = \hat{\mathbf{K}}_{d \to f}(\tau)\hat{g}_{d,\tau}$

and $(\hat{\mathbf{K}}_{f\to d}(\cdot)\hat{g}_f)_\tau = \hat{\mathbf{K}}_{f\to d}(\tau)\hat{g}_{f,\tau}$, and where the restrictions $\hat{\mathcal{B}}_d(t,T,E,\hat{g}_d)$ and $\hat{\mathcal{B}}_f(t,T,E,\hat{g}_f)$ to the binomial trees Γ_d^n and Γ_f^n of the values $\mathcal{B}_d(t,T,E,g_d)$ and $\mathcal{B}_f(t,T,E,g_f)$ of Bermudan options in the binomial market environments $\mathbf{V}_d^B$ and $\mathbf{V}_f^B$ are defined by

(3.21)
$$\hat{\mathcal{B}}_d(t,T,E,Q_d(\cdot)g_d) = Q_d(t)\mathcal{B}_d(t,T,E,g_d), \quad g_d : E \to \Pi_+,$$
$$\hat{\mathcal{B}}_f(t,T,E,Q_f(\cdot)g_f) = Q_f(t)\mathcal{B}_d(t,T,E,g_f), \quad g_f : E \to \Pi_+,$$

with $(Q_d(\cdot)g_d)_\tau = Q_d(\tau)g_{d,\tau}$ and $(Q_f(\cdot)g_f)_\tau = Q_f(\tau)g_{f,\tau}$.

In the particular case of Bermudan call and put options valued in the binomial market environments $\mathbf{V}_d^B$ and $\mathbf{V}_f^B$ restricted to the binomial trees Γ_d and Γ_f, that is, in the restricted binomial market environments $\hat{\mathbf{V}}_d^B$ and $\hat{\mathbf{V}}_f^B$, the symmetry relationships in (3.20) take the form:

(3.22)
$$\hat{C}_d(t,T,S,X) = SX\,\hat{P}_f(t,T,1/S,1/X),$$
$$\hat{P}_d(t,T,S,X) = SX\,\hat{C}_f(t,T,1/S,1/X),$$

(3.23)
$$\hat{C}_f(t,T,S',X') = S'X'\,\hat{P}_d(t,T,1/S',1/X'),$$
$$\hat{P}_f(t,T,S',X') = S'X'\,\hat{C}_d(t,T,1/S',1/X'),$$

where S is in $\Gamma_d^n(t)$ and S' is in $\Gamma_f^n(t)$.

We shall postpone describing the numerical significance of the symmetry relationships (3.20) until after we present the symmetry in the foreign exchange market for American options with general time-dependent payoffs. This is because it is the values of foreign exchange American options in the Black and Scholes market environment, rather than Bermudan options in the Black and Scholes market environment, that are numerically approximated by the values of foreign exchange Bermudan options in the binomial market environment.

3.5   Symmetry for American Options

Now we present the symmetry in the foreign exchange market for American options with general time-dependent payoffs.

Assume that for t and T in the trading time set $\mathcal{T}$ with $t \leq T$, the interval $[t,T]$ is contained in $\mathcal{T}$.

An *American option* (on a general underlying security) with inception time t, expiration time T, and with a (time-dependent) payoff $g : [t,T] \to \Pi_+$ is a contract that gives the right, but not the obligation, to receive the payoff $g_\tau(S_\tau)$ at any time τ in the interval $[t, T]$, where the price of the underlying security at time τ is S_τ.

Among American options, one of the most widely traded are American call and put options.

An *American call option* and an *American put option* are American options with time-independent payoffs $(\cdot - X)^+ : [t, T] \to \Pi_+$ and $(X - \cdot)^+ : [t, T] \to \Pi_+$, where $X > 0$ is called the strike price. Financially, American call and put options give the right but not the obligation to by or to sell one unit of an underlying security at the strike price at any time during the life of the option.

The payoffs $(\cdot - X)^+ : [t, T] \to \Pi_+$ and $(X - \cdot)^+ : [t, T] \to \Pi_+$ are called a *call payoff* and a *put payoff* respectively.

We comment again that although the call and put payoffs defined for European, Bermudan and American options are mathematically different functions, financially they have the same meaning. That is why, whenever ambiguity is not likely, we use the same notations for these call and put payoffs.

In the foreign exchange market in which the underlying security is the domestic or foreign currency, we denote the value of the American option in currency d on currency f with inception time t, expiration time T, and with payoff g_d in currency d by $\mathcal{A}_d(t, T, g_d) = \mathcal{A}_d(t, T, g_d)(S_t)$. Similarly, we denote the value of the American option in currency f on currency d with inception time t, expiration time T, and with payoff g_f in currency f by $\mathcal{A}_f(t, T, g_f) = \mathcal{A}_f(t, T, g_f)(S_t')$. Clearly $\mathcal{A}_d(t, T, g_d)$ and $\mathcal{A}_f(t, T, g_f)$ are in Π_+.

In this notation, the values of the foreign exchange American call and put options can be represented as

$$C_d(t, T, S, X) = \mathcal{A}_d(t, T, (\cdot - X)^+)(S),$$
$$P_d(t, T, S, X) = \mathcal{A}_d(t, T, (X - \cdot)^+)(S),$$

and

$$C_f(t, T, S', X') = \mathcal{A}_f(t, T, (\cdot - X')^+)(S'),$$
$$P_f(t, T, S', X') = \mathcal{A}_f(t, T, (X' - \cdot)^+)(S').$$

As for the preceding cases of European and Bermudan options, it is possible to give a coordinate-free, that is, a (domestic-foreign) basis-independent, description of an American option in the foreign exchange market in terms of an abstract American option.

An *abstract American option* with inception time t, expiration time T and with a (time-dependent) abstract payoff $g : [t, T] \to \Pi_+$ is a contract that gives the right, but not the obligation, to receive the abstract payoff g_τ in $\mathbf{\Pi}_+$ at any time τ in the interval $[t, T]$.

It is clear that the value $\mathcal{A}(t, T, g)$ of an abstract American option is an element of $\mathbf{\Pi}$, even of $\mathbf{\Pi}_+$. Therefore, the expression of $\mathcal{A}(t, T, g)$, as an element of $\mathbf{\Pi}$, in the domestic basis is the value $\mathcal{A}_d(t, T, g_d)$ in Π_+ of the American option in currency d on currency f with inception time t, expiration time T, and with payoff g_d, where g_d is the expression of the abstract payoff g in the domestic basis. On the other hand, the expression of $\mathcal{A}(t, T, g)$, as an element of $\mathbf{\Pi}$, in the foreign basis is the value $\mathcal{A}_f(t, T, g_f)$ in Π_+ of the American option in currency f on currency d with inception time t, expiration time T, and with payoff g_f, where g_f is the expression of the abstract payoff g in the foreign basis.

This is the symmetry we are seeking for foreign exchange American options. This symmetry associates financially equivalent American options on the opposite sides of the foreign exchange market, and states that foreign exchange American options on the opposite sides of the market are financially equivalent provided that their payoffs are financially equivalent.

In more detail, the value $\mathcal{A}_f(t, T, g_f)$, converted into currency d, of the American option in currency f on currency d with payoff g_f has to be equal to the value $\mathcal{A}_d(t, T, g_d)$ of the American option in currency d on currency f with payoff g_d, where g_f is the payoff g_d converted into currency f. Otherwise there must be an arbitrage opportunity in the foreign exchange market. Similarly, the value $\mathcal{A}_d(t, T, g_d)$, converted into currency f, of the American option in currency d on currency f with payoff g_d has to be equal to the value $\mathcal{A}_f(t, T, g_f)$ of the American option in currency f on currency d with payoff g_f, where g_d is the payoff g_f converted into currency d. In symbols these symmetry relationships for foreign exchange American options can be stated as

$$(3.24) \quad \begin{aligned} \mathcal{A}_d(t, T, g_d) &= \mathbf{K}\mathcal{A}_f(t, T, \mathbf{K}g_d), \quad g_d : [t, T] \to \Pi_+, \\ \mathcal{A}_f(t, T, g_f) &= \mathbf{K}\mathcal{A}_d(t, T, \mathbf{K}g_f), \quad g_f : [t, T] \to \Pi_+. \end{aligned}$$

In the particular case of American call and put options, the symmetry relationships (3.24) take the form of the relationships (1.1) and (1.2).

The symmetry relationships (3.24) hold for foreign exchange American options valued in one of the major cases of market environments, the Black and Scholes market environments $\mathbf{V}_d^{BS}$ and $\mathbf{V}_f^{BS}$. We note that American options do not exist in the other major case of the market environment, the binomial market environment, since these options require the trading time sets to contain an entire time interval but the trading time sets in the binomial market environment are discrete.

However, it is well known that for sufficiently large n, the values of foreign exchange Bermudan options in the binomial market environments $\mathbf{V}_d^{B_n}$ and $\mathbf{V}_f^{B_n}$ restricted to the binomial trees Γ_d^n and Γ_f^n, that is, in the restricted binomial market environments $\hat{\mathbf{V}}_d^{B_n}$ and $\hat{\mathbf{V}}_f^{B_n}$ approximate the values of the corresponding American options in the Black and Scholes market environment. See, for example, Jarrow and Rudd [11]. Therefore, (3.20) indicates the fact that the symmetry relationships hold at all levels of accuracy in the numerical evaluation of foreign exchange American options in the Black and Scholes market environment via such approximations. In Section 8.7 we will present a detailed analysis of the practical significance of this fact using the example of such numerical evaluation of foreign exchange American call and put options.

Finally we comment that the concepts of an abstract payoff, an abstract evolution operator, and an abstract European, Bermudan and American option allow for a coordinate-free, that is, (domestic-foreign) basis-independent description of financial phenomena in a foreign exchange market.

This completes the presentation of symmetry relationships for payoffs and for European, Bermudan, and American options in a foreign exchange market. Further symmetry relationships, based on these, are presented in the next two chapters.

Chapter 4

Further Symmetries

The reason why we presented the symmetry relationships for foreign exchange European, Bermudan and American options with general payoffs is not just for the sake of complete generality. Financially it is clear that for any foreign exchange option, no matter how complex it is, there exists a financially equivalent option on the opposite side of the foreign exchange market. What is not clear in every given instance, is how to find such a financially equivalent option explicitly. In this regard, the preceding generality allows us to obtain the explicit form of the symmetry relationships for one of the major classes of exotic options, namely barrier options. In turn, we consider barrier options in the generality of an arbitrary time-dependent barrier which can be activated not only on the entire life of the option, but on any discrete set of times during its life. Furthermore, we consider barrier options of European, Bermudan and American styles with the underlying European, Bermudan and American options having general and, in the latter two cases, time-dependent payoffs. As was shown by Kholodnyi in [18], such barrier options can be viewed as Bermudan options with certain payoffs for discretely activated barriers or as American options with certain payoffs for continuously activated barriers. This fact enables us to use the symmetry relationships presented in Chapter 3 to derive the symmetry relationships for such barrier options.

In this chapter, we also illustrate the power of the machinery we have developed by derivation of the symmetry relationships for the Greek letters, in particular for deltas and gammas, of arbitrary foreign exchange options. The chapter ends with presentation of the symmetry relationships for foreign exchange European, Bermudan and American call and put options in the particular case of a market environment which we call the exchange-rate homogeneous

55

market environment.

Once again, in order to maintain the financial clarity of the exposition, we postpone mathematical proofs until Chapter 9.

4.1   Symmetry for Barrier Options

In this section we present the symmetry relationships for one of the major classes of exotic options, namely barrier options. There are two levels of generality in which we consider barrier options. First, we consider barrier options with arbitrary time-dependent barriers which can be activated not only on the entire life of the option, but on any discrete set of times during its life. Such barrier options with partial barriers are becoming more and more popular, especially in the foreign exchange market [5], providing a flexible alternative to relatively expensive underlying options and relatively cheap barrier options with continuously activated barriers. Second, we consider barrier options of European, Bermudan and American styles with the underlying European, Bermudan and American options having general and, in the latter two cases, time-dependent payoffs. The derivation of the symmetry relationships for such barrier options is based on the fact shown in [18], that such barrier options can be viewed as Bermudan options with certain payoffs for discretely activated barriers or as American options with certain payoffs for continuously activated barriers. In this way the symmetry relationships for such barrier options are essentially a particular case of the symmetry relationships for Bermudan and American options presented in Chapter 3.

Assume that for t and T in the trading time set $\mathcal{T}$ with $t \leq T$, the *barrier activation set* E_b is either an interval of the form $[t, T]$ in $\mathcal{T}$ or a discrete subset of the form $\{t'_i : i = 0, 1, \ldots, n'\}$ with $t \leq t'_0 < t'_1 < \ldots < t'_{n'-1} < t'_{n'} \leq T$ in $\mathcal{T}$. Henceforth, if the barrier activation set is discrete, we will assume without loss of generality that it contains T.

A *down and in barrier option* of a European style, Bermudan style, or American style (on a general underlying security) with inception time t, expiration time T, barrier activation set E_b, and with a (time-dependent) barrier $b : E_b \to \mathbb{R}_+$ is a contract specified as follows: (1) if there exists a time τ in the set E_b such that $S_\tau \leq b(\tau)$, then at the first such time τ the barrier option becomes, respectively, the European option, the Bermudan option, or the American option with inception time τ, expiration time T and payoff g, (2) otherwise, the option expires worthless.

The definition of an *up and in barrier option* of European, Bermudan, or American style is completely analogous except that the inequality $S_\tau \leq b(\tau)$

in condition (1) is replaced with the inequality $S_\tau \geq b(\tau)$.

We shall refer to such down and in or up and in barrier options of a European, Bermudan, or American style as down and in or up and in barrier options *on* European, Bermudan, or American options with payoff g, respectively.

A *down and out barrier option* of a European style, Bermudan style or American style (on a general underlying security) with inception time t, expiration time T, barrier activation set E_b and with a (time-dependent) barrier $b : E_b \rightarrow \mathbb{R}_+$ is a portfolio consisting of a long position on a European, Bermudan or American option with inception time t, expiration time T, and payoff g, and a short position on a down and in barrier option with inception time t, expiration time T, barrier activation set E_b, and barrier $b : E_b \rightarrow \mathbb{R}_+$ on the European, Bermudan, or American option with payoff g, respectively.

The definition of an *up and out barrier option* of European, Bermudan, or American style is completely analogous except that the short position on the down and in barrier option is replaced by a short position *on* the up and in barrier option of either European, Bermudan, or American style.

Again, we shall refer to such down and out or up and out barrier options of a European, Bermudan, or American style as down and out or up and out barrier options *on* European, Bermudan, or American options with payoff g, respectively.

In order to be consistent with the definitions of Bermudan and American options, in these definitions of barrier options we assumed that the trading time set $\mathcal{T}$ contains an exercise time set E with $t \leq t_0 < t_1 < ... < t_{n-1} < t_n = T$ when the barrier option is of Bermudan style and contains the interval $[t, T]$ when the barrier option is of American style.

For convenience, we introduce a uniform notation for foreign exchange barrier options of European, Bermudan and American styles.

Denote the value of the down and in barrier option with inception time t, expiration time T, barrier activation set E_b, and barrier $b_d : E_b \rightarrow \mathbb{R}_+$ in currency d on currency f by

$$DI_d(t, T, E_b, b_d, E_\mathcal{O}, g_d) = DI_d(t, T, E_b, b_d, E_\mathcal{O}, g_d)(S_t),$$

where it is on the European option if $E_\mathcal{O} = \{T\}$, or on the Bermudan option if $E_\mathcal{O} = E$, or on the American option if $E_\mathcal{O} = [t, T]$, each with payoff g_d in currency d. Similarly, denote the value of the down and in barrier option with inception time t, expiration time T, barrier activation set E_b, and barrier $b_f : E_b \rightarrow \mathbb{R}_+$ in currency f on currency d by

$$DI_f(t, T, E_b, b_f, E_\mathcal{O}, g_f) = DI_f(t, T, E_b, b_f, E_\mathcal{O}, g_f)(S'_t).$$

Clearly $DI_d(t, T, E_b, b_d, E_\mathcal{O}, g_d)$ and $DI_f(t, T, E_b, b_f, E_\mathcal{O}, g_f)$ are in Π_+.

Denote the value of the up and in barrier option with inception time t, expiration time T, barrier activation set E_b, and barrier $b_d : E_b \to \mathbb{R}_+$ in currency d on currency f by

$$UI_d(t, T, E_b, b_d, E_\mathcal{O}, g_d) = UI_d(t, T, E_b, b_d, E_\mathcal{O}, g_d)(S_t),$$

where it is on the European option if $E_\mathcal{O} = \{T\}$, or on the Bermudan option if $E_\mathcal{O} = E$, or on the American option if $E_\mathcal{O} = [t, T]$, each with payoff g_d in currency d. Similarly, we denote the value of the up and in barrier option with inception time t, expiration time T, barrier activation set E_b, and barrier $b_f : E_b \to \mathbb{R}_+$ in currency f on currency d by

$$UI_f(t, T, E_b, b_f, E_\mathcal{O}, g_f) = UI_f(t, T, E_b, b_f, E_\mathcal{O}, g_f)(S'_t).$$

Clearly $UI_d(t, T, E_b, b_d, E_\mathcal{O}, g_d)$ and $UI_f(t, T, E_b, b_f, E_\mathcal{O}, g_f)$ are in Π_+.

Denote the value of the down and out barrier option with inception time t, expiration time T, barrier activation set E_b, and barrier $b_d : E_b \to \mathbb{R}_+$ in currency d on currency f by

$$DO_d(t, T, E_b, b_d, E_\mathcal{O}, g_d) = DO_d(t, T, E_b, b_d, E_\mathcal{O}, g_d)(S_t),$$

where it is on the European option if $E_\mathcal{O} = \{T\}$, or on the Bermudan option if $E_\mathcal{O} = E$, or on the American option if $E_\mathcal{O} = [t, T]$, each with payoff g_d in currency d. Similarly, we denote the value of the down and out barrier option with inception time t, expiration time T, barrier activation set E_b, and barrier $b_f : E_b \to \mathbb{R}_+$ in currency f on currency d by

$$DO_f(t, T, E_b, b_f, E_\mathcal{O}, g_f) = DO_f(t, T, E_b, b_f, E_\mathcal{O}, g_f)(S'_t).$$

Clearly $DO_d(t, T, E_b, b_d, E_\mathcal{O}, g_d)$ and $DO_f(t, T, E_b, b_f, E_\mathcal{O}, g_f)$ are in Π_+.

Denote the value of the up and out barrier option with inception time t, expiration time T, barrier activation set E_b, and barrier $b_d : E_b \to \mathbb{R}_+$ in currency d on currency f by

$$UO_d(t, T, E_b, b_d, E_\mathcal{O}, g_d) = UO_d(t, T, E_b, b_d, E_\mathcal{O}, g_d)(S_t),$$

where it is on the European option if $E_\mathcal{O} = \{T\}$, or on the Bermudan option if $E_\mathcal{O} = E$, or on the American option if $E_\mathcal{O} = [t, T]$, each with payoff g_d in currency d. Similarly, we denote the value of the up and out barrier option with inception time t, expiration time T, barrier activation set E_b, and barrier $b_f : E_b \to \mathbb{R}_+$ in currency f on currency d by

$$UO_f(t, T, E_b, b_f, E_\mathcal{O}, g_f) = UO_f(t, T, E_b, b_f, E_\mathcal{O}, g_f)(S'_t).$$

Clearly $UO_d(t, T, E_b, b_d, E_\mathcal{O}, g_d)$ and $UO_f(t, T, E_b, b_f, E_\mathcal{O}, g_f)$ are in Π_+.

In symbols, the values of down and out and up and out barrier options in currency d on currency f are given by

(4.1)
$$DO_d(t, T, E_b, b_d, E_\mathcal{O}, g_d) = \mathcal{O}_d(t, T, E_\mathcal{O}, g_d) - DI_d(t, T, E_b, b_d, E_\mathcal{O}, g_d)$$
$$UO_d(t, T, E_b, b_d, E_\mathcal{O}, g_d) = \mathcal{O}_d(t, T, E_\mathcal{O}, g_d) - UI_d(t, T, E_b, b_d, E_\mathcal{O}, g_d),$$

where, for convenience, we introduce the uniform notation $\mathcal{O}_d(t, T, E_\mathcal{O}, g_d)$ for the values of European, Bermudan and American options in currency d on currency f:

(4.2)
$$\mathcal{O}_d(t, T, E_\mathcal{O}, g_d) = \begin{cases} \mathcal{E}_d(t, T, g_d) & \text{if } E_\mathcal{O} = \{T\}, \\ \mathcal{B}_d(t, T, E, g_d) & \text{if } E_\mathcal{O} = E, \\ \mathcal{A}_d(t, T, g_d) & \text{if } E_\mathcal{O} = [t, T]. \end{cases}$$

Similarly, in symbols the values of down and out and up and out barrier options in currency f on currency d are given by

(4.3)
$$DO_f(t, T, E_b, b_f, E_\mathcal{O}, g_f) = \mathcal{O}_f(t, T, E_\mathcal{O}, g_f) - DI_f(t, T, E_b, b_f, E_\mathcal{O}, g_f)$$
$$UO_f(t, T, E_b, b_f, E_\mathcal{O}, g_f) = \mathcal{O}_f(t, T, E_\mathcal{O}, g_f) - UI_f(t, T, E_b, b_f, E_\mathcal{O}, g_f),$$

where again we introduce the uniform notation $\mathcal{O}_f(t, T, E_\mathcal{O}, g_f)$ for the values of European, Bermudan and American options in currency f on currency d:

(4.4)
$$\mathcal{O}_f(t, T, E_\mathcal{O}, g_f) = \begin{cases} \mathcal{E}_f(t, T, g_f) & \text{if } E_\mathcal{O} = \{T\}, \\ \mathcal{B}_f(t, T, E, g_f) & \text{if } E_\mathcal{O} = E, \\ \mathcal{A}_f(t, T, g_f) & \text{if } E_\mathcal{O} = [t, T]. \end{cases}$$

Note that we introduced the uniform notation for barrier options of European, Bermudan and American styles, and for European, Bermudan and American options not just for the sake of convenience. This uniform notation, as a set function of the argument $E_\mathcal{O}$ with domain $\{\{T\}, E, [0, T]\}$ and range in the set of all possible options (more precisely, in the set of the values of all possible options) illustrates in seed form the concept of a *universal contingent claim* as a *multiplicative measure* given by the product integral of, generally speaking, nonlinear and noncommutative operators. The concept of a universal contingent claim provides a unified framework for the analysis of a very general class of financial derivatives. For example, Bermudan and American options represent one type of a universal contingent claim in the sense that their values in the product integral representations (8.1) and (8.5) have the same operators, the maximum operators, in addition to the evolution operators $\mathbf{V}_d(t,T)$ or $\mathbf{V}_f(t,T)$. This universal contingent claim has the value $\mathcal{O}_d(t,T,E_{\mathcal{O}},g_d)$ or $\mathcal{O}_f(t,T,E_{\mathcal{O}},g_f)$ with the exercise time set $E_{\mathcal{O}}$ that could be of a Bermudan type E or of an American type $[0,T]$. We comment that as a relevant example we have used Bermudan and American options in the foreign exchange setting. The concepts of universal contingent claim and of multiplicative measure were introduced and studied by Kholodnyi in [22], [21], [15] and [16] and will be briefly illustrated again later in Chapter 8. The symmetry relationships for foreign exchange universal contingent claims were established by Kholodnyi in [25] and are beyond the scope of this book.

Denote by H the Heaviside function

$$H(x) = \begin{cases} 1 & \text{if } x \geq 0, \\ 0 & \text{if } x < 0. \end{cases}$$

A down and in barrier option can be viewed [18] as a Bermudan or American option with a particular class of payoffs:

(4.5)

$$DI_d(t,T,E_b,b_d,E_{\mathcal{O}},g_d) = \begin{cases} \mathcal{B}_d(t,T,E_b,g_d^D) & \text{if } E_b = E, \\ \mathcal{A}_d(t,T,g_d^D) & \text{if } E_b = [t,T], \end{cases}$$

where the time-dependent payoff g_d^D is given by

(4.6) $\qquad g_{d,\tau}^D(S_\tau) = H(b_d(\tau) - S_\tau)\mathcal{O}_d(\tau,T,E_{\mathcal{O}},g_d)(S_\tau), \quad \tau \in E_b;$

and

(4.7)

$$DI_f(t,T,E_b,b_f,E_{\mathcal{O}},g_f) = \begin{cases} \mathcal{B}_f(t,T,E_b,g_f^D) & \text{if } E_b = E, \\ \mathcal{A}_f(t,T,g_f^D) & \text{if } E_b = [t,T], \end{cases}$$

where the time-dependent payoff g_f^D is given by

(4.8) $\qquad g_{f,\tau}^D(S_\tau') = H(b_f(\tau) - S_\tau')\mathcal{O}_f(\tau,T,E_{\mathcal{O}},g_f)(S_\tau'), \quad \tau \in E_b.$

Similarly, an up and in barrier option can be viewed [18] as a Bermudan or American option with a particular class of payoffs:

(4.9)
$$UI_d(t, T, E_b, b_d, E_{\mathcal{O}}, g_d) = \begin{cases} \mathcal{B}_d(t, T, E_b, g_d^U) & \text{if } E_b = E, \\ \mathcal{A}_d(t, T, g_d^U) & \text{if } E_b = [t, T], \end{cases}$$

where the time-dependent payoff g_d^U is given by

(4.10)
$$g_{d,\tau}^U(S_\tau) = H(S_\tau - b_d(\tau))\mathcal{O}_d(\tau, T, E_{\mathcal{O}}, g_d)(S_\tau), \quad \tau \in E_b;$$

and

(4.11)
$$UI_f(t, T, E_b, b_f, E_{\mathcal{O}}, g_f) = \begin{cases} \mathcal{B}_f(t, T, E_b, g_f^U) & \text{if } E_b = E, \\ \mathcal{A}_f(t, T, g_f^U) & \text{if } E_b = [t, T], \end{cases}$$

where the time-dependent payoff g_f^U is given by

(4.12)
$$g_{f,\tau}^U(S_\tau') = H(S_\tau' - b_f(\tau))\mathcal{O}_f(\tau, T, E_{\mathcal{O}}, g_f)(S_\tau'), \quad \tau \in E_b.$$

The symmetry relationships for Bermudan and American options combine with the preceding representations of down and in and up and in barrier options as Bermudan or American options and with the relationships (4.1) and (4.3) to imply the following symmetry relationships for barrier options:

Down and in barrier options

(4.13)
$$DI_d(t, T, E_b, b_d, E_{\mathcal{O}}, g_d) = \mathbf{K}UI_f(t, T, E_b, 1/b_d, E_{\mathcal{O}}, \mathbf{K}g_d), \quad g_d : E_{\mathcal{O}} \to \Pi_+,$$
$$DI_f(t, T, E_b, b_f, E_{\mathcal{O}}, g_f) = \mathbf{K}UI_d(t, T, E_b, 1/b_f, E_{\mathcal{O}}, \mathbf{K}g_f), \quad g_f : E_{\mathcal{O}} \to \Pi_+,$$

Up and in barrier options

(4.14)
$$UI_d(t, T, E_b, b_d, E_{\mathcal{O}}, g_d) = \mathbf{K}DI_f(t, T, E_b, 1/b_d, E_{\mathcal{O}}, \mathbf{K}g_d), \quad g_d : E_{\mathcal{O}} \to \Pi_+,$$
$$UI_f(t, T, E_b, b_f, E_{\mathcal{O}}, g_f) = \mathbf{K}DI_d(t, T, E_b, 1/b_f, E_{\mathcal{O}}, \mathbf{K}g_f), \quad g_f : E_{\mathcal{O}} \to \Pi_+,$$

Down and out barrier options

(4.15)
$$DO_d(t, T, E_b, b_d, E_{\mathcal{O}}, g_d) = \mathbf{K}UO_f(t, T, E_b, 1/b_d, E_{\mathcal{O}}, \mathbf{K}g_d), \quad g_d : E_{\mathcal{O}} \to \Pi_+,$$
$$DO_f(t, T, E_b, b_f, E_{\mathcal{O}}, g_f) = \mathbf{K}UO_d(t, T, E_b, 1/b_f, E_{\mathcal{O}}, \mathbf{K}g_f), \quad g_f : E_{\mathcal{O}} \to \Pi_+,$$

Up and out barrier options

(4.16)
$$UO_d(t, T, E_b, b_d, E_{\mathcal{O}}, g_d) = \mathbf{K}DO_f(t, T, E_b, 1/b_d, E_{\mathcal{O}}, \mathbf{K}g_d), \quad g_d : E_{\mathcal{O}} \to \Pi_+,$$
$$UO_f(t, T, E_b, b_f, E_{\mathcal{O}}, g_f) = \mathbf{K}DO_d(t, T, E_b, 1/b_f, E_{\mathcal{O}}, \mathbf{K}g_f), \quad g_f : E_{\mathcal{O}} \to \Pi_+.$$

In the particular case where the barrier options are on European, Bermudan, or American call or put options, the preceding symmetry relationships for barrier options take the form

Down and in call and put barrier options

(4.17)
$$DIC_d(t, T, E_b, b_d, E_{\mathcal{O}}, S, X) = SX\, UIP_f(t, T, E_b, 1/b_d, E_{\mathcal{O}}, 1/S, 1/X),$$
$$DIC_f(t, T, E_b, b_f, E_{\mathcal{O}}, S', X') = S'X'\, UIP_d(t, T, E_b, 1/b_f, E_{\mathcal{O}}, 1/S', 1/X'),$$

(4.18)
$$DIP_d(t, T, E_b, b_d, E_{\mathcal{O}}, S, X) = SX\, UIC_f(t, T, E_b, 1/b_d, E_{\mathcal{O}}, 1/S, 1/X),$$
$$DIP_f(t, T, E_b, b_f, E_{\mathcal{O}}, S', X') = S'X'\, UIC_d(t, T, E_b, 1/b_f, E_{\mathcal{O}}, 1/S', 1/X').$$

Up and in call and put barrier options

(4.19)
$$UIC_d(t, T, E_b, b_d, E_{\mathcal{O}}, S, X) = SX\, DIP_f(t, T, E_b, 1/b_d, E_{\mathcal{O}}, 1/S, 1/X),$$
$$UIC_f(t, T, E_b, b_f, E_{\mathcal{O}}, S', X') = S'X'\, DIP_d(t, T, E_b, 1/b_f, E_{\mathcal{O}}, 1/S', 1/X'),$$

(4.20)
$$UIP_d(t, T, E_b, b_d, E_{\mathcal{O}}, S, X) = SX\, DIC_f(t, T, E_b, 1/b_d, E_{\mathcal{O}}, 1/S, 1/X),$$
$$UIP_f(t, T, E_b, b_f, E_{\mathcal{O}}, S', X') = S'X'\, DIC_d(t, T, E_b, 1/b_f, E_{\mathcal{O}}, 1/S', 1/X').$$

Down and out call and put barrier options

(4.21)
$$DOC_d(t,T,E_b,b_d,E_\mathcal{O},S,X) = SX\,UOP_f(t,T,E_b,1/b_d,E_\mathcal{O},1/S,1/X),$$
$$DOC_f(t,T,E_b,b_f,E_\mathcal{O},S',X') = S'X'\,UOP_d(t,T,E_b,1/b_f,E_\mathcal{O},1/S',1/X'),$$

(4.22)
$$DOP_d(t,T,E_b,b_d,E_\mathcal{O},S,X) = SX\,UOC_f(t,T,E_b,1/b_d,E_\mathcal{O},1/S,1/X),$$
$$DOP_f(t,T,E_b,b_f,E_\mathcal{O},S',X') = S'X'\,UOC_d(t,T,E_b,1/b_f,E_\mathcal{O},1/S',1/X').$$

Up and out call and put barrier options

(4.23)
$$UOC_d(t,T,E_b,b_d,E_\mathcal{O},S,X) = SX\,DOP_f(t,T,E_b,1/b_d,E_\mathcal{O},1/S,1/X),$$
$$UOC_f(t,T,E_b,b_f,E_\mathcal{O},S',X') = S'X'\,DOP_d(t,T,E_b,1/b_f,E_\mathcal{O},1/S',1/X'),$$

(4.24)
$$UOP_d(t,T,E_b,b_d,E_\mathcal{O},S,X) = SX\,DOC_f(t,T,E_b,1/b_d,E_\mathcal{O},1/S,1/X),$$
$$UOP_f(t,T,E_b,b_f,E_\mathcal{O},S',X') = S'X'\,DOC_d(t,T,E_b,1/b_f,E_\mathcal{O},1/S',1/X').$$

The notation in the preceding symmetry relationships (4.17)–(4.24) is as follows:

(4.25)
$$DIC_d(t,T,E_b,b_d,E_\mathcal{O},S,X) = DI_d(t,T,E_b,b_d,E_\mathcal{O},(\cdot - X)^+)(S),$$
$$DIC_f(t,T,E_b,b_f,E_\mathcal{O},S',X') = DI_f(t,T,E_b,b_f,E_\mathcal{O},(\cdot - X')^+)(S'),$$
$$DIP_d(t,T,E_b,b_d,E_\mathcal{O},S,X) = DI_d(t,T,E_b,b_d,E_\mathcal{O},(X - \cdot)^+)(S),$$
$$DIP_f(t,T,E_b,b_f,E_\mathcal{O},S',X') = DI_f(t,T,E_b,b_f,E_\mathcal{O},(X' - \cdot)^+)(S'),$$

(4.26)
$$UIC_d(t,T,E_b,b_d,E_\mathcal{O},S,X) = UI_d(t,T,E_b,b_d,E_\mathcal{O},(\cdot - X)^+)(S),$$
$$UIC_f(t,T,E_b,b_f,E_\mathcal{O},S',X') = UI_f(t,T,E_b,b_f,E_\mathcal{O},(\cdot - X')^+)(S'),$$
$$UIP_d(t,T,E_b,b_d,E_\mathcal{O},S,X) = UI_d(t,T,E_b,b_d,E_\mathcal{O},(X - \cdot)^+)(S),$$
$$UIP_f(t,T,E_b,b_f,E_\mathcal{O},S',X') = UI_f(t,T,E_b,b_f,E_\mathcal{O},(X' - \cdot)^+)(S'),$$

(4.27)
$$DOC_d(t, T, E_b, b_d, E_\mathcal{O}, S, X) = DO_d(t, T, E_b, b_d, E_\mathcal{O}, (\cdot - X)^+)(S),$$
$$DOC_f(t, T, E_b, b_f, E_\mathcal{O}, S', X') = DO_f(t, T, E_b, b_f, E_\mathcal{O}, (\cdot - X')^+)(S'),$$
$$DOP_d(t, T, E_b, b_d, E_\mathcal{O}, S, X) = DO_d(t, T, E_b, b_d, E_\mathcal{O}, (X - \cdot)^+)(S),$$
$$DOP_f(t, T, E_b, b_f, E_\mathcal{O}, S', X') = DO_f(t, T, E_b, b_f, E_\mathcal{O}, (X' - \cdot)^+)(S'),$$

(4.28)
$$UOC_d(t, T, E_b, b_d, E_\mathcal{O}, S, X) = UO_d(t, T, E_b, b_d, E_\mathcal{O}, (\cdot - X)^+)(S),$$
$$UOC_f(t, T, E_b, b_f, E_\mathcal{O}, S', X') = UO_f(t, T, E_b, b_f, E_\mathcal{O}, (\cdot - X')^+)(S'),$$
$$UOP_d(t, T, E_b, b_d, E_\mathcal{O}, S, X) = UO_d(t, T, E_b, b_d, E_\mathcal{O}, (X - \cdot)^+)(S),$$
$$UOP_f(t, T, E_b, b_f, E_\mathcal{O}, S', X') = UO_f(t, T, E_b, b_f, E_\mathcal{O}, (X' - \cdot)^+)(S').$$

This completes the presentation of the symmetry relationships for barrier options. The mathematical proofs can be found in Chapter 9. In the next section we present the symmetry relations for Greek letters, in particular for delta and gamma of arbitrary foreign exchange options.

4.2 Symmetry for Greek Letters

The symmetry relationships for foreign exchange options have their counterparts for the 'Greek' letters of the options. The interesting nontrivial cases are those Greek letters that require partial differentiation with respect to the exchange rate, namely delta and gamma. We note, however, that the use of these Greek letters for the sensitivity analysis of an option or portfolio containing options is not valid in a general market environment. For details on this issue see [13].

Let $\Phi_d = \Phi_d(S)$ denote the value of an arbitrary foreign exchange option in currency d on currency f at any time from a trading time set. Similarly, let $\Phi_f = \Phi_f(S')$ denote the value of an arbitrary foreign exchange option in currency f on currency d at the same time. Moreover, let Φ_d, as a payoff in currency d, be the payoff Φ_f in currency f, converted to the currency d and vice versa, that is,

(4.29)
$$\Phi_d(S) = (\mathbf{K}\Phi_f)(S) = S\Phi_f(1/S),$$
$$\Phi_f(S') = (\mathbf{K}\Phi_d)(S') = S'\Phi_d(1/S').$$

Denote by $\Delta_d(S)$ and $\Delta_f(S')$ the deltas of the foreign exchange options with values Φ_d and Φ_f, that is, the first partial derivatives of $\Phi_d(S)$ and $\Phi_f(S')$ with respect to the exchange rates S and S'. Similarly, denote by $\Gamma_d(S)$ and $\Gamma_f(S')$ the gammas of the foreign exchange options with values Φ_d and Φ_f, that is, the second partial derivatives of $\Phi_d(S)$ and $\Phi_f(S')$ with respect to the exchange rates S and S'.

By straightforward differentiation of the equalities in (4.29) with respect to the exchange rates, we obtain the symmetry relationships for delta and gammas of foreign exchange options:

$$\Delta_d(S) = \Phi_f(1/S) - (1/S)\Delta_f(1/S),$$
$$\Delta_f(S') = \Phi_d(1/S') - (1/S')\Delta_d(1/S'),$$

and

$$\Gamma_d(S) = (1/S^3)\Gamma_f(1/S),$$
$$\Gamma_f(S') = (1/S'^3)\Gamma_d(1/S').$$

In the particular cases of European, Bermudan, and American call and put options in currency d on currency f, by letting

$$\Phi_d(S) = C_d(t,T,S,X) \quad \text{and} \quad \Phi_f(S') = X\,P_f(t,T,S',1/X)$$

and, in turn,

$$\Phi_d(S) = P_d(t,T,S,X) \quad \text{and} \quad \Phi_f(S') = X\,C_f(t,T,S',1/X),$$

these symmetry relationships take the form

$$\Delta_d^c(S,X) = X\,P_f(t,T,1/S,1/X) - (X/S)\,\Delta_f^p(1/S,1/X),$$
$$\Delta_d^p(S,X) = X\,C_f(t,T,1/S,1/X) - (X/S)\,\Delta_f^c(1/S,1/X),$$

and

$$\Gamma_d^c(S,X) = (1/S^3)\Gamma_f^p(1/S,1/X),$$
$$\Gamma_d^p(S,X) = (1/S^3)\Gamma_f^c(1/S,1/X),$$

where the superscripts c and p indicate call and put options.

Similarly, in the particular cases of European, Bermudan, and American call and put options in currency f on currency d, by letting

$$\Phi_f(S') = C_f(t, T, S', X') \quad \text{and} \quad \Phi_d(S) = X' P_d(t, T, S, 1/X')$$

and, in turn,

$$\Phi_f(S') = P_f(t, T, S', X') \quad \text{and} \quad \Phi_d(S) = X' C_d(t, T, S, 1/X'),$$

these symmetry relationships take the form

$$\Delta_f^c(S', X') = X' P_d(t, T, 1/S', 1/X') - (X'/S') \Delta_d^p(1/S', 1/X'),$$
$$\Delta_f^p(S', X') = X' C_d(t, T, 1/S', 1/X') - (X'/S') \Delta_d^c(1/S', 1/X'),$$

and

$$\Gamma_f^c(S', X') = (1/S'^3)\Gamma_d^p(1/S', 1/X'),$$
$$\Gamma_f^p(S', X') = (1/S'^3)\Gamma_d^c(1/S', 1/X').$$

This completes the presentation of the symmetry relationships for the Greek letters of foreign exchange options. In the next section we present the symmetry relationships for foreign exchange European, Bermudan and American call and put options in the particular case of a market environment which we call the exchange-rate homogeneous market environment.

4.3 Symmetry in the Exchange-Rate Homogeneous Market Environment

In this section we consider symmetry relationships for European, Bermudan, and American call and put options in a foreign exchange market in which the values of these options are homogeneous functions of degree one jointly in the exchange rates and in the strike prices:

(4.30)
$$\lambda C_d(t, T, S, X) = C_d(t, T, \lambda S, \lambda X),$$
$$\lambda P_d(t, T, S, X) = P_d(t, T, \lambda S, \lambda X),$$

(4.31)
$$\lambda C_f(t, T, S', X') = C_f(t, T, \lambda S', \lambda X'),$$
$$\lambda P_f(t, T, S', X') = P_f(t, T, \lambda S', \lambda X'),$$

for all nonnegative λ.

This is the case, for example, if a model of the market place is chosen that treats the dynamics of the exchange rates as a stochastic process in which the probability distribution of the return per unit of domestic currency invested in the foreign currency is independent of the level of the exchange rate between the domestic and foreign currencies. (For more details, see Merton [30], Theorems 6 and 9, and Appendix A.) In this case the return per unit of foreign currency invested in the domestic currency is clearly independent of the level of the exchange rate between the foreign and domestic currencies. Such stochastic processes and such foreign exchange market environments we will call *exchange-rate homogeneous*.

For a definition of the exchange-rate homogeneous market environment based on the evolution operators $\mathbf{V}_d(t,T)$ and $\mathbf{V}_f(t,T)$, and for the symmetry relationships for general European, Bermudan and American options along with barrier options in such a market environment based on the Kelvin transform $\mathbf{K}_c$ with arbirary $c > 0$ see [24].

Note that the homogeneity relationships (4.30) and (4.31) are not just rescaling relationships due to the choice of new units of domestic or foreign currency in the same state of the foreign exchange market as expressed by the exchange rates. Rather, the relationships in (4.30) assert that the portfolio of λ call or put options in currency d on currency f with strike X which are evaluated at the state of the market S, as expressed by the exchange rate between currency d and currency f has the same value as the call or put option in currency d on currency f with strike λX which are evaluated at the state of the market λS, as expressed by the exchange rate between currency d and currency f. Similarly, the relationships in (4.31) assert that the portfolio of λ call or put options in currency f on currency d with strike X' which are evaluated at the state of the market S', as expressed by the exchange rate between currency f and currency d has the same value as the call and put option in currency f on currency d with strike $\lambda X'$ which are evaluated at the state of the market $\lambda S'$, as expressed by the exchange rate between currency f and currency d.

One of the major examples of exchange-rate homogeneous market environments are the binomial market environments $\mathbf{V}_d^{B_n}$ and $\mathbf{V}_f^{B_n}$, and the Black and Scholes market environments $\mathbf{V}_d^{BS}$ and $\mathbf{V}_f^{BS}$.

As an illustration we prove here the exchange-rate homogeneity of the Black and Scholes market environments $\mathbf{V}_d^{BS}$ and $\mathbf{V}_f^{BS}$ directly from the well known explicit expressions (2.22) and (2.23) for the values of European call and put options given by the Black and Scholes formulas. More precisely, we show that the homogeneity relationships (4.30) and (4.31) hold for the foreign exchange European call and put options in the Black and Scholes market environments $\mathbf{V}_d^{BS}$ and $\mathbf{V}_f^{BS}$, that is

$$\lambda C_d^{BS}(t,T,S,X) = C_d^{BS}(t,T,\lambda S,\lambda X),$$
$$\lambda P_d^{BS}(t,T,S,X) = P_d^{BS}(t,T,\lambda S,\lambda X),$$
$$\lambda C_f^{BS}(t,T,S',X') = C_f^{BS}(t,T,\lambda S',\lambda X'),$$
$$\lambda P_f^{BS}(t,T,S',X') = P_f^{BS}(t,T,\lambda S',\lambda X'),$$

for all nonnegative λ.

We prove only the first of the preceding homogeneity relationships, since the proof of the remaining relationships is completely analogous. The proof in the case of positive λ follows from the following chain of equalities:

$$\lambda C_d^{BS}(t,T,S,X)$$
$$= \lambda \left(e^{-r_f(T-t)} S\, N\left(\frac{\log(S/X) + (r_d - r_f + \frac{1}{2}\sigma^2)(T-t)}{\sigma\sqrt{T-t}}\right) \right.$$
$$\left. - e^{-r_d(T-t)} X\, N\left(\frac{\log(S/X) + (r_d - r_f - \frac{1}{2}\sigma^2)(T-t)}{\sigma\sqrt{T-t}}\right) \right)$$
$$= e^{-r_f(T-t)} (\lambda S)\, N\left(\frac{\log((\lambda S)/(\lambda X)) + (r_d - r_f + \frac{1}{2}\sigma^2)(T-t)}{\sigma\sqrt{T-t}}\right)$$
$$- e^{-r_d(T-t)} (\lambda X)\, N\left(\frac{\log((\lambda S)/(\lambda X)) + (r_d - r_f - \frac{1}{2}\sigma^2)(T-t)}{\sigma\sqrt{T-t}}\right)$$
$$= C_d^{BS}(t,T,\lambda S,\lambda X),$$

where the first and the last equalities are due to the expression (2.22), and the second equality is due to trivial algebra. The case of λ equal to zero is understood as the corresponding limit.

In an exchange-rate homogeneous foreign exchange market environment, the symmetry relationships (1.1) and (1.2) for European, Bermudan and American call and put options can be rewritten as

$$\text{(4.32)} \qquad \begin{aligned} C_d(t,T,S,X) &= P_f(t,T,X,S), \\ P_d(t,T,S,X) &= C_f(t,T,X,S), \end{aligned}$$

$$\text{(4.33)} \qquad \begin{aligned} C_f(t,T,S',X') &= P_d(t,T,X',S'), \\ P_f(t,T,S',X') &= C_d(t,T,X',S'). \end{aligned}$$

The relationships in (4.32) assert that the call and put options in currency d on currency f are evaluated at the state of the market S, as expressed by the exchange rate between currency d and currency f, while the call and put options in currency f on currency d are evaluated at the state of the market X, as expressed by the exchange rate between currency f and currency d. Similarly, the relationships in (4.33) assert that the call and put options in currency f on currency d are evaluated at the state of the market S', as expressed by the exchange rate between currency f and currency d, while the call and put options in currency d on currency f are evaluated at the state of the market X', as expressed by the exchange rate between currency d and currency f.

More precisely, as an example, the first relationship in (4.32) asserts that if at time t the state of the foreign exchange market as expressed by the exchange rate between currency d and currency f is S, then the value $C_d(t,T,S,X)$ is the same as the value $P_f(t,T,X,S)$ if at time t the state of the foreign exchange market as expressed by the exchange rate between currency f and currency d is X. From a practical standpoint, unless $X = 1/S$, this relationship cannot be observed in a foreign exchange market. The obstacle is that the foreign exchange market at any time t cannot be in two different states as expressed by the exchange rate S between currency d and currency f and by the exchange rate X between currency f and currency d. Therefore the values $C_d(t,T,S,X)$ and $P_f(t,T,X,S)$ cannot be observed simultaneously. Furthermore, even if at time t the foreign exchange market is in the state S as expressed by the exchange rate between currency d and currency f and at time t' the foreign exchange market is in the state X as expressed by the exchange rate between currency f and currency d, the values $C_d(t,T,S,X)$ and $P_f(t',T',X,S)$, with equal times to maturity $T - t = T' - t'$ cannot have any meaningful association although they can each be separately observed.

Nevertheless, despite the impossibility of observing the symmetry relationships (4.32) and (4.33) in an exchange-rate homogeneous foreign exchange market environment, they can be interpreted as symmetry relationships in the chosen model of a foreign exchange market environment.

On the other hand, if the values of call and put options in the symmetry relationships (4.32) and (4.33) are evaluated in a time-homogeneous market environment, that is, if they are just functions of the time to maturity $T - t$ then these symmetry relationships can actually be observed in the foreign exchange market. For example, the first symmetry relationship in (4.32) can be observed as follows. For a given time to maturity $T - t = T' - t'$, if the state of the foreign exchange market at time t as expressed by the exchange rate between currency d and currency f is S, the market value $C_d(t,T,S,X)$

has to be equal to the market value $P_f(t', T', X, S)$ if the state of the foreign exchange market at time t' as expressed by the exchange rate between currency f and currency d is X. Moreover, if any of the symmetry relationships (4.32) and (4.33) are observed in the foreign exchange market for all traded values of times to maturity and for all traded values of the exchange rates and strike prices, then this is an experimental indication that the foreign exchange market environment is time-homogeneous and exchange-rate homogeneous.

We have seen in this Chapter how the symmetry relationships for payoffs and for European, Bermudan, and American options with general payoffs in a foreign exchange market can be used to derive symmetry relationships for one of the major classes of exotic options, namely barrier options; for the Greek letters of arbitrary foreign exchange options; and for foreign exchange European, Bermudan and American call and put options in the particular case of a market environment, the exchange-rate homogeneous market environment. In the next Chapter, we present an example of how the symmetry relationships for payoffs and for European, Bermudan, and American options with general payoffs can be used to guide the choice of new payoffs that smooth the corners of the payoffs of the European, Bermudan, and American call and put options.

Chapter 5

Options with Consistently Smoothed Payoffs

As we indicated in the previous Chapter, the reason why we presented the symmetry relationships for foreign exchange European, Bermudan and American options with arbitrary payoffs is not just for the sake of complete generality. In this Chapter we present another reason for doing so. It is well known that there is an advantage to smooth the corners of payoffs of European, Bermudan, and American options such as European, Bermudan, and American call and put options so that even for the case of these call and put options we are forced to consider payoffs of a general type. The practical reason for this smoothing is that the resulting options with smoothed payoffs are easier to hedge, for example, in the Black and Scholes market environment. At the same time, these smoothed payoffs, and hence the values of these resulting options, are close to those of the original options with unsmoothed payoffs. We use the symmetry relationships for European, Bermudan, and American options with general payoffs to guide the choice of new payoffs that smooth the corners for the payoffs of European, Bermudan, and American call and put options so that the form of the symmetry relationships for these call and put options is maintained for the resulting options with the smoothed payoffs.

5.1 The Importance of Smooth Payoffs

It is well known that it is an advantage to smooth the corners of payoffs of European, Bermudan, and American options such as European, Bermudan, and American call and put options. The practical reason for this smoothing is that the resulting options with smoothed payoffs are easier to hedge, for example, in the Black and Scholes type of market environment. The difficulty with hedging, for example, European, Bermudan, and American call and put options that are near the money is that their deltas are very sensitive to movements in the exchange rate, particularly as the options near their times of expiration. In other words, they have high gammas. The reason for this high sensitivity is that payoffs for European, Bermudan, and American call and put options have corners, that is, discontinuous first derivatives when the underlying exchange rate equals the strike. As a result, the second derivatives of the payoffs are delta functions concentrated at the strike prices. Since the values of the options as they near their times of expiration converge to their payoffs at expiration, the options' gammas, being second derivatives of the values of the options, converge to these delta functions.

This problem can be removed by smoothing the payoffs appropriately. The question then arises is how best to smooth the payoffs. There are clearly many ways one could approximate a payoff function with a corner by a smooth function. Natural requirements, for example, could be that a smoothed payoff differ from the original payoff with corner on as small an interval as we choose and by as small an amount as we choose. Hart and Ross proposed in [10] particular smoothed payoffs for foreign exchange European call and put options that are both financially reasonable and satisfy these requirements. On the other hand, it is also natural to require, in addition, that the smoothed payoffs be taken on both sides of the foreign exchange market and be related in the same way as the original call and put payoffs, that is, that the form of the symmetry relationships for the call and put payoffs be maintained for the resulting smoothed payoffs. As it turns out, the smoothed payoffs suggested by Hart and Ross do not satisfy this additional symmetry requirement.

In this chapter we first present Hart and Ross's smoothed payoffs and show that they do not satisfy the symmetry requirement. Then we present new smoothed payoffs for call and put payoffs that have all the desirable properties possessed by Hart and Ross's smoothed payoffs, but also satisfy the symmetry relationships that resemble those for call and put payoffs.

This new smoothing applies not only to European, but also to Bermudan, and American call and put options. A crucial role is played by the fact that the symmetry relationships for foreign exchange European, Bermudan, and American options are obtained in the generality of arbitrary payoffs rather than just for call and put payoffs. It is this generality that enables us to handle the symmetry relationships for these smoothed payoffs which are of a general type and not just of the form of call and put payoffs, as well as the symmetry relationships for European, Bermudan, and American options with these smoothed payoffs.

For the general statement, and a solution, of the problem of approximating two arbitrary financially equivalent payoffs and two financially equivalent European, Bermudan and American options with arbitrary payoffs by respectively financially equivalent smooth payoffs and financially equivalent European, Bermudan and American options with smooth payoffs see [23].

Once again we postpone mathematical proofs until Chapter 10 in order to maintain the financial clarity of our presentation.

5.2   The Continuous Strike Range Call and Put Options

Hart and Ross discuss in [10] the hedging difficulty for European options with payoffs that have corners and proposed that, instead of European call options in currency d on currency f with strike X and in currency f on currency d with strike X', it is better to consider European options with the smoothed payoffs $h_d^c(L_d, U_d, X)$ and $h_f^c(L_f, U_f, X')$ defined by

(5.1)
$$h_d^c(L_d, U_d, X)(S_T) = \left([(S_T - L_d)^+]^2 - [(S_T - U_d)^+]^2\right)/2(U_d - L_d),$$
$$h_f^c(L_f, U_f, X')(S_T') = \left([(S_T' - L_f)^+]^2 - [(S_T' - U_f)^+]^2\right)/2(U_f - L_f),$$

where $L_d < X < U_d$ with $X = (L_d + U_d)/2$, and $L_f < X' < U_f$ with $X' = (L_f + U_f)/2$.

Figure 5.1 gives a typical example of the payoff $h_d^c(X(1 - \epsilon), X(1 + \epsilon), X)$ with $\epsilon = 0.1$ compared to the payoff of a European call option with strike $X = 100$.

The payoffs $h_d^c(L_d, U_d, X)$ and $h_f^c(L_f, U_f, X')$ have the following three properties:

1. $h_d^c(L_d, U_d, X)(S_T)$ equals the value of the call payoff $(S_T - X)^+$ for S_T outside the interval (L_d, U_d) and $h_f^c(L_f, U_f, X')(S_T')$ equals the value of the call payoff $(S_T' - X')^+$ for S_T' outside the interval (L_f, U_f),

2. $h_d^c(L_d, U_d, X)$ and $h_f^c(L_f, U_f, X')$ have continuous first derivatives with respect to the exchange rates, and

3. $h_d^c(L_d, U_d, X)$ and $h_f^c(L_f, U_f, X')$ converge to the call payoffs $(\cdot - X)^+$ and $(\cdot - X')^+$ in various natural topologies, for example, pointwise, as

$\epsilon \to 0$, where $L_d = X(1 - \epsilon)$, $U_d = X(1 + \epsilon)$ and $L_f = X'(1 - \epsilon)$, $U_f = X'(1 + \epsilon)$, for $\epsilon > 0$.

In foreign exchange market environments in which the operators $\mathbf{V}_d(t, T)$ and $\mathbf{V}_f(t, T)$ are continuous with respect to the topologies in Property 3, the values $\mathcal{E}_d(t, T, h_d^c(L_d, U_d, X))$ and $\mathcal{E}_f(t, T, h_f^c(L_f, U_f, X'))$ of European options with payoffs $h_d^c(L_d, U_d, X)$ and $h_f^c(L_f, U_f, X')$ converge to the values $C_d(t, T, \cdot, X)$ and $C_f(t, T, \cdot, X')$ of the standard European call options. For example, in the foreign exchange binomial market environment and the Black and Scholes market environment, this convergence is at least pointwise, that is, $\mathcal{E}_d(t, T, h_d^c(L_d, U_d, X))(S_t)$ and $\mathcal{E}_f(t, T, h_f^c(L_f, U_f, X'))(S_t')$ converge to $C_d(t, T, S_t, X)$ and $C_f(t, T, S_t', X')$ for any given S_t and S_t'. In [10], Hart and Ross refer to the European options with values $\mathcal{E}_d(t, T, h_d^c(L_d, U_d, X))$ and $\mathcal{E}_f(t, T, h_f^c(L_f, U_f, X'))$ as *continuous strike range (European call) options*.

Although Hart and Ross did not discuss in [10] smoothing for put payoffs, according to their approach, instead of European put options in currency d on currency f with strike X and in currency f on currency d with strike X', we consider European options with the smoothed payoffs $h_d^p(L_d, U_d, X)$ and $h_f^p(L_f, U_f, X')$ defined by

(5.2)

$$h_d^p(L_d, U_d, X)(S_T) = \left([(U_d - S_T)^+]^2 - [(L_d - S_T)^+]^2\right)/2(U_d - L_d),$$
$$h_f^p(L_f, U_f, X')(S_T') = \left([(U_f - S_T')^+]^2 - [(L_f - S_T')^+]^2\right)/2(U_f - L_f),$$

where $L_d < X < U_d$ with $X = (L_d + U_d)/2$, and $L_f < X' < U_f$ with $X' = (L_f + U_f)/2$.

The payoffs $h_d^p(L_d, U_d, X)$ and $h_f^p(L_f, U_f, X')$ have properties analogous to the properties of $h_d^c(L_d, U_d, X)$ and $h_f^c(L_f, U_f, X')$:

1. $h_d^p(L_d, U_d, X)(S_T)$ equals the value of the put payoff $(X - S_T)^+$ for S_T outside the interval (L_d, U_d) and $h_f^p(L_f, U_f, X')(S_T')$ equals the value of the put payoff $(X' - S_T')^+$ for S_T' outside the interval (L_f, U_f),

2. $h_d^p(L_d, U_d, X)$ and $h_f^p(L_f, U_f, X')$ have continuous first derivatives with respect to the exchange rates, and

3. $h_d^p(L_d, U_d, X)$ and $h_f^p(L_f, U_f, X')$ converge to the put payoffs $(X - \cdot)^+$ and $(X' - \cdot)^+$ in various natural topologies, for example, pointwise, as $\epsilon \to 0$, where $L_d = X(1 - \epsilon)$, $U_d = X(1 + \epsilon)$ and $L_f = X'(1 - \epsilon)$, $U_f = X'(1 + \epsilon)$, for $\epsilon > 0$.

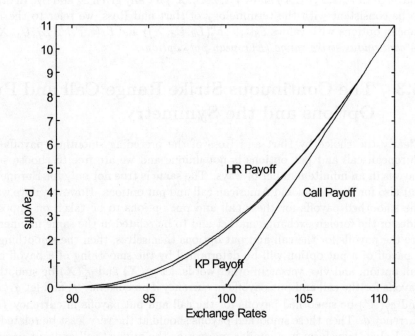

Figure 5.1: **Comparison of a call payoff with a continuous strike range call payoff and a consistently smoothed call payoff.** Comparison of the payoff for a standard call option with strike $X = 100$ (Call payoff), the continuous strike range call payoff $h_d^c(L_d, U_d, X)$ with $L_d = X(1-\epsilon)$ and $U_d = X(1+\epsilon)$ suggested by Hart and Ross (HR Payoff), and the payoff $g_d^c(L_d, U_d, X)$ with $L_d = X/(1+\epsilon)$ and $U_d = X(1+\epsilon)$ described in the text (KP Payoff), where $X = 100$ and $\epsilon = 0.1$.

Once again, in foreign exchange market environments in which the operators $\mathbf{V}_d(t,T)$ and $\mathbf{V}_f(t,T)$ are continuous with respect to the topologies in Property 3, the values $\mathcal{E}_d(t, T, h_d^p(L_d, U_d, X))$ and $\mathcal{E}_f(t, T, h_f^p(L_f, U_f, X'))$ of European options with payoffs $h_d^p(L_d, U_d, X)$ and $h_f^p(L_f, U_f, X')$ converge to the values $P_d(t, T, \cdot, X)$ and $P_f(t, T, \cdot, X')$ of the standard European put options. For example, in the foreign exchange binomial market environment and the Black and Scholes market environment, this convergence is at least pointwise, that is, $\mathcal{E}_d(t, T, h_d^p(L_d, U_d, X))(S_t)$ and $\mathcal{E}_f(t, T, h_f^p(L_f, U_f, X'))(S_t')$

converge to $P_d(t, T, S_t, X)$ and $P_f(t, T, S'_t, X')$ for any given S_t and S'_t. In order to be consistent with the terminology of Hart and Ross, we refer to the European options with values $\mathcal{E}_d(t, T, h^p_d(L_d, U_d, X))$ and $\mathcal{E}_f(t, T, h^p_f(L_f, U_f, X'))$ as *continuous strike range (European put) options*.

5.3   The Continuous Strike Range Call and Put Options and the Symmetry

Clearly the choice by Hart and Ross of the preceding smoothed payoffs for European call and put options is not unique and we are free to choose such payoffs in an infinite number of ways. The same is true not only for European, but also for Bermudan and American call and put options. However, if we want the smoothed payoffs for these call and put options to be taken on opposite sides of the foreign exchange market and to be related in the same manner as are the payoffs for the call and put options themselves, then the smoothing of a payoff of a put option will be determined by the smoothing of a payoff of a call option, and vice versa. In other words, let $g^c_d(X)$ and $g^p_d(X)$ be smoothed payoffs for the call and put payoffs in currency d on currency f, and let $g^c_f(X')$ and $g^p_f(X')$ be smoothed payoffs for the call and put payoffs in currency f on currency d. Then these smoothed payoffs should at the very least be related by the Kelvin transform in a similar manner as the original call and put payoffs. In symbols, we should have

(5.3)
$$\mathbf{K}g^c_d(X) = X\, g^p_f(1/X),$$
$$\mathbf{K}g^p_d(X) = X\, g^c_f(1/X),$$

(5.4)
$$\mathbf{K}g^c_f(X') = X'\, g^p_d(1/X'),$$
$$\mathbf{K}g^p_f(X') = X'\, g^c_d(1/X').$$

However, the Kelvin transforms of the payoffs

$$h^c_d(L_d, U_d, X) \quad \text{and} \quad h^p_d(L_d, U_d, X)$$

are not of the form

$$h^p_f(L_f, U_f, X') \quad \text{and} \quad h^c_f(L_f, U_f, X'),$$

and vice versa, so that for any choice of L_d, U_d, L_f and U_f, the symmetry relationships (5.3) and (5.4) are never satisfied.

5.4 The Consistently Smoothed Call and Put Payoffs and the Symmetry

We now present new smoothed payoffs for call and put payoffs which, unlike the payoffs of Hart and Ross, are consistent with the symmetry relationships for call and put payoffs. Define the smoothed payoffs $g_d^c(L_d, U_d, X)$ and $g_d^p(L_d, U_d, X)$ for the call and put payoffs $(\cdot - X)^+$ and $(X - \cdot)^+$ in currency d on currency f by

$$g_d^c(L_d, U_d, X)(S) = \begin{cases} 0 & \text{for } S \le L_d, \\ \beta_d(S - L_d)^{2\alpha_d+1}/S^{\alpha_d} & \text{for } L_d < S < U_d, \\ S - X & \text{for } S \ge U_d, \end{cases}$$

$$g_d^p(L_d, U_d, X)(S) = \begin{cases} X - S & \text{for } S \le L_d, \\ \delta_d(U_d - S)^{2\gamma_d+1}/S^{\gamma_d} & \text{for } L_d < S < U_d, \\ 0 & S \ge U_d, \end{cases}$$

where $0 < L_d < U_d$,

$$X = X(L_d, U_d) = \sqrt{L_d U_d},$$

$$\alpha_d = \alpha_d(L_d, U_d) = \frac{U_d}{L_d + U_d} \frac{X - L_d}{U_d - X},$$

$$\beta_d = \beta_d(L_d, U_d) = \frac{U_d^{\alpha_d}(U_d - X)}{(U_d - L_d)^{2\alpha_d+1}},$$

$$\gamma_d = \gamma_d(L_d, U_d) = \frac{L_d}{L_d + U_d} \frac{U_d - X}{X - L_d},$$

$$\delta_d = \delta_d(L_d, U_d) = \frac{L_d^{\gamma_d}(X - L_d)}{(U_d - L_d)^{2\gamma_d+1}}.$$

We note that $L_d < X < U_d$ and α_d, β_d, γ_d and δ_d are positive, since $0 < L_d < U_d$.

We call the payoffs $g_d^c(L_d, U_d, X)$ and $g_d^p(L_d, U_d, X)$ *consistently smoothed call payoff* and *consistently smoothed put payoff* in currency d on currency f.

Similarly, for the opposite side of the foreign exchange market, define the smoothed payoffs $g_f^c(L_f, U_f, X')$ and $g_f^p(L_f, U_f, X')$ for the call and put payoffs $(\cdot - X')^+$ and $(X' - \cdot)^+$ in currency f on currency d by

$$g_f^c(L_f, U_f, X')(S') = \begin{cases} 0 & \text{for } S' \le L_f, \\ \beta_f(S' - L_f)^{2\alpha_f+1}/S'^{\alpha_f} & \text{for } L_f < S' < U_f, \\ S' - X' & \text{for } S' \ge U_f, \end{cases}$$

$$g_f^p(L_f, U_f, X')(S') = \begin{cases} X' - S' & \text{for } S' \leq L_f, \\ \delta_f(U_f - S')^{2\gamma_f+1}/S'^{\gamma_f} & \text{for } L_f < S' < U_f, \\ 0 & S' \geq U_f, \end{cases}$$

where $0 < L_f < U_f$,

$$X' = X'(L_f, U_f) = \sqrt{L_f U_f},$$

$$\alpha_f = \alpha_f(L_f, U_f) = \frac{U_f}{L_f + U_f} \frac{X' - L_f}{U_f - X'},$$

$$\beta_f = \beta_f(L_f, U_f) = \frac{U_f^{\alpha_f}(U_f - X')}{(U_f - L_f)^{2\alpha_f+1}},$$

$$\gamma_f = \gamma_f(L_f, U_f) = \frac{L_f}{L_f + U_f} \frac{U_f - X'}{X' - L_f},$$

$$\delta_f = \delta_f(L_f, U_f) = \frac{L_f^{\gamma_f}(X' - L_f)}{(U_f - L_f)^{2\gamma_f+1}}.$$

We note that $L_f < X < U_f$ and α_f, β_f, γ_f and δ_f are positive, since $0 < L_f < U_f$.

We call the payoffs $g_f^c(L_f, U_f, X')$ and $g_f^p(L_f, U_f, X')$ *consistently smoothed call payoff* and *consistently smoothed put payoff* in currency f on currency d.

Figure 5.1 gives a typical example of the payoff $g_d^c(X/(1+\epsilon), X(1+\epsilon), X)$ compared to the call payoff with strike $X = 100$ and compared to the continuous strike range call payoff of Hart and Ross $h_d^c(X(1-\epsilon), X(1+\epsilon), X)$, where $\epsilon = 0.1$.

The payoffs $g_d^c(L_d, U_d, X)$ and $g_f^c(L_f, U_f, X')$ have the following three properties, which are analogous to the above properties of the payoffs $h_d^c(L_d, U_d, X)$ and $h_f^c(L_f, U_f, X')$:

1. $g_d^c(L_d, U_d, X)(S_T)$ equals the value of the call payoff $(S_T - X)^+$ for S_T outside the interval (L_d, U_d) and $g_f^c(L_f, U_f, X')(S_T')$ equals the value of the call payoff $(S_T' - X')^+$ for S_T' outside the interval (L_f, U_f),

2. $g_d^c(L_d, U_d, X)$ and $g_f^c(L_f, U_f, X')$ have continuous first derivatives with respect to the exchange rates, and

3. $g_d^c(L_d, U_d, X)$ and $g_f^c(L_f, U_f, X')$ converge to the call payoffs $(\cdot - X)^+$ and $(\cdot - X')^+$ in various natural topologies, for example, pointwise, as $\epsilon \to 0$, where $L_d = X/(1+\epsilon)$, $U_d = X(1+\epsilon)$ and $L_f = X'/(1+\epsilon)$, $U_f = X'(1+\epsilon)$, for $\epsilon > 0$.

Denote by $C_d(t, T, S, X, L_d, U_d)$ and $C_f(t, T, S', X', L_f, U_f)$ the values

$$\mathcal{O}_d(t, T, E_{\mathcal{O}}, g_d^c(L_d, U_d, X))(S) \quad \text{and} \quad \mathcal{O}_f(t, T, E_{\mathcal{O}}, g_f^c(L_f, U_f, X'))(S')$$

of European, Bermudan and American options with payoffs

$$g_d^c(L_d, U_d, X) \quad \text{and} \quad g_f^c(L_f, U_f, X'),$$

where the uniform notation $\mathcal{O}_d(t, T, E_{\mathcal{O}}, g_d)$ and $\mathcal{O}_f(t, T, E_{\mathcal{O}}, g_f)$ for European, Bermudan and American options was defined in Chapter 4

In foreign exchange market environments in which the values

$$\mathcal{O}_d(t, T, E_{\mathcal{O}}, g_d) \quad \text{and} \quad \mathcal{O}_f(t, T, E_{\mathcal{O}}, g_f)$$

are continuous functions of the payoffs g_d and g_f with respect to the topologies in the preceding Property 3, the values

$$\mathcal{O}_d(t, T, E_{\mathcal{O}}, g_d^c(L_d, U_d, X)) \quad \text{and} \quad \mathcal{O}_f(t, T, E_{\mathcal{O}}, g_f^c(L_f, U_f, X'))$$

converge to the values $C_d(t, T, \cdot, X)$ and $C_f(t, T, \cdot, X')$ of the standard European, Bermudan, and American call options, respectively.

Turning to the consistently smoothed put payoffs, we see that the payoffs $g_d^p(L_d, U_d, X)$ and $g_f^p(L_f, U_f, X')$ have the following three properties analogous to the properties of $h_d^p(L_d, U_d, X)$ and $h_f^p(L_f, U_f, X')$:

1. $g_d^p(L_d, U_d, X)(S_T)$ equals the value of the put payoff $(X - S_T)^+$ for S_T outside the interval (L_d, U_d) and $g_f^p(L_f, U_f, X')(S_T')$ equals the value of the put payoff $(X' - S_T')^+$ for S_T' outside the interval (L_f, U_f),

2. $g_d^p(L_d, U_d, X)$ and $g_f^p(L_f, U_f, X')$ have continuous first derivatives with respect to the exchange rates, and

3. $g_d^p(L_d, U_d, X)$ and $g_f^p(L_f, U_f, X')$ converge to the put payoffs $(X - \cdot)^+$ and $(X' - \cdot)^+$ in various natural topologies, for example, pointwise, as $\epsilon \to 0$, where $L_d = X/(1 + \epsilon)$, $U_d = X(1 + \epsilon)$ and $L_f = X'/(1 + \epsilon)$, $U_f = X'(1 + \epsilon)$, for $\epsilon > 0$.

Denote by $P_d(t, T, S, X, L_d, U_d)$ and $P_f(t, T, S', X', L_f, U_f)$ the values

$$\mathcal{O}_d(t, T, E_{\mathcal{O}}, g_d^p(L_d, U_d, X))(S) \quad \text{and} \quad \mathcal{O}_f(t, T, E_{\mathcal{O}}, g_f^p(L_f, U_f, X'))(S')$$

of European, Bermudan and American options with payoffs

$$g_d^p(L_d, U_d, X) \quad \text{and} \quad g_f^p(L_f, U_f, X').$$

In foreign exchange market environments in which the values

$$\mathcal{O}_d(t, T, E_{\mathcal{O}}, g_d) \quad \text{and} \quad \mathcal{O}_f(t, T, E_{\mathcal{O}}, g_f)$$

are continuous functions of the payoffs g_d and g_f with respect to the topologies in the preceding Property 3, the values

$$\mathcal{O}_d(t, T, E_{\mathcal{O}}, g_d^p(L_d, U_d, X)) \quad \text{and} \quad \mathcal{O}_f(t, T, E_{\mathcal{O}}, g_f^p(L_f, U_f, X'))$$

converge to the values $P_d(t, T, \cdot, X)$ and $P_f(t, T, \cdot, X')$ of the standard European, Bermudan, and American put options, respectively.

5.5   Symmetry for Options with the Consistently Smoothed Call and Put Payoffs

In addition to the properties presented in the Section 5.4, the four pairs of the consistently smoothed call and put payoffs $g_d^c(L_d, U_d, X)$ and $g_f^p(L_f, U_f, X')$, $g_d^p(L_d, U_d, X)$ and $g_f^c(L_f, U_f, X')$, $g_f^c(L_f, U_f, X')$ and $g_d^p(L_d, U_d, X)$, and $g_f^p(L_f, U_f, X')$ and $g_d^c(L_d, U_d, X)$ are related via the Kelvin transform as in (5.3) and (5.4). That is, the form of the symmetry relationships for call and put payoffs is maintained by the symmetry relationships for these the consistently smoothed call and put payoffs. In symbols:

$$(5.5) \quad \begin{aligned} \mathbf{K} g_d^c(L_d, U_d, X) &= X\, g_f^p(1/U_d, 1/L_d, 1/X), \\ \mathbf{K} g_d^p(L_d, U_d, X) &= X\, g_f^c(1/U_d, 1/L_d, 1/X), \end{aligned}$$

$$(5.6) \quad \begin{aligned} \mathbf{K} g_f^c(L_f, U_f, X') &= X'\, g_d^p(1/U_f, 1/L_f, 1/X'), \\ \mathbf{K} g_f^p(L_f, U_f, X') &= X'\, g_d^c(1/U_f, 1/L_f, 1/X'). \end{aligned}$$

The practical significance of the symmetry relationships in (5.5) and (5.6) for the consistently smoothed call and put payoffs $g_d^c(L_d, U_d, X)$, $g_d^p(L_d, U_d, X)$, $g_f^c(L_f, U_f, X')$, and $g_f^p(L_f, U_f, X')$ is that the values of the European, Bermudan, and American options with these payoffs satisfy the following symmetry relationships:

$$(5.7) \quad \begin{aligned} C_d(t, T, S, X, L_d, U_d) &= SX\, P_f(t, T, 1/S, 1/X, 1/U_d, 1/L_d), \\ P_d(t, T, S, X, L_d, U_d) &= SX\, C_f(t, T, 1/S, 1/X, 1/U_d, 1/L_d), \end{aligned}$$

and

(5.8)
$$C_f(t,T,S',X',L_f,U_f) = S'X' P_d(t,T,1/S',1/X',1/U_f,1/L_f),$$
$$P_f(t,T,S',X',L_f,U_f) = S'X' C_d(t,T,1/S',1/X',1/U_f,1/L_f).$$

It is clear that the form of the symmetry relationships for European, Bermudan and American call and put options (1.1) and (1.2) is maintained by the symmetry relationships (5.7) and (5.8) for European, Bermudan and American options with consistently smoothed call and put payoffs.

In the next Chapter we give an overview of the main applications of the symmetry relationships in a foreign exchange market described earlier in the book.

Chapter 6

Applications

In this Chapter we present the main applications of the symmetry relationships in a foreign exchange market introduced in the previous Chapters.

6.1 Detecting Arbitrage

The first application is that the symmetry relationships provide a means for detecting a new type of true arbitrage in the foreign exchange option market directly from the market data. If, within the interval arising from bid-ask spreads and transaction costs, any of the symmetry relationships is observed to be false, then an arbitrage profit is assured. We stress that, due to the general nature of the symmetry relationships, these arbitrage opportunities are true or absolute, that is, they depend only upon observed option premiums and are not conditional on any particular choice of a model of the foreign exchange option market.

6.2 Selecting a Financially Equivalent Portfolio

The symmetry relationships also provide a means of selecting, for a given foreign exchange portfolio, which may include options, a financially equivalent portfolio on the opposite side of a foreign exchange market, when for some reason it is not convenient or not possible to achieve the original portfolio. This is, for example, the case when one needs to take an option position and the foreign exchange option market is one-sided in the sense that the desired option is not listed on the official exchanges on the desired side of the market but its financially equivalent option is listed on such exchanges on the opposite side of the market.

6.3 Detecting Inconsistent Option Pricing Models

At the same time, the symmetry relationships provide a screen to help in detecting inconsistent models of foreign exchange option markets. For example, for European options, at the very least the evolution operators $V_d(t, T)$ and $V_f(t, T)$ for such a model must satisfy the symmetry relationships in (3.4). As we saw in Section 3.2, this is the case for such major instances of foreign exchange market environments as the binomial market environment and the Black and Scholes market environment.

6.4 Instant Valuation of Foreign Counterparts

Another practical outcome of the symmetry relationships is that one can instantly value and analyze the foreign counterpart of a domestic foreign exchange option or portfolio of such options without lengthy additional re-calculations. This instant valuation and analysis is crucially important because these options are essential and major components of most modern financial strategies, both individually and as a key ingredient of large financial portfolios. These strategies have to be constantly monitored and adjusted with the consequence that the values of the component options have to be continually re-evaluated in real time based on the current value of the strategy and on its future objectives. Such real-time re-evaluation may soon be more than just a preferable choice. In Germany, for example, new regulations imply that this is a legal requirement [3].

In the standard Black and Scholes market environment, having computed the value of a European option on one side of the foreign exchange market, the computational time required to re-compute its counterpart on the opposite side of the market is comparable to the computational time needed to compute the counterpart by applying the symmetry relationship to the initial European option. That is, there may be no real savings of computational time. However, in more realistic models of foreign exchange markets currently in use, which include features such as domestic and foreign interest rates and volatilities that depend on time and exchange rate, even the computation of the values of European options requires computational time that is, by itself, prohibitively large. In this case, having computed the value of a European option on one side of the foreign exchange market, the computational time required to re-compute its counterpart on the opposite side of the market is no longer comparable with the computational time needed to compute the counterpart by applying the symmetry relationship to the initial European option which is still essentially instantaneous.

The advantage of applying the symmetry relationships to value the counterpart on the opposite side of the market of a given foreign exchange option dramatically increases for more complex options than European options such as Bermudan, American, barrier, and other exotic options. This advantage is significant even in the standard Black and Scholes market environment. But once more realistic models of the market place are invoked which include features such as domestic and foreign interest rates and volatilities that depend on time and exchange rate, use of the symmetry relationships becomes crucial and possibly the only practical way to proceed.

6.5   Instant Evaluation of Greek Letters

The discussion in the previous section is valid not only for the part of the analysis of foreign exchange option portfolios based just on the values of the component options and their counterparts on the opposite side of the market, but also for more sophisticated analyses, such as sensitivity analysis, that require, in addition, the evaluation of their Greek letters. The interesting non-trivial cases are those Greek letters that require partial differentiation with respect to the exchange rate, namely delta and gamma.

6.6   Approximation Schemes

From the practical standpoint, in order to compute the values of foreign exchange options that do not have closed-form solutions, which includes almost all options, some approximation scheme has to be employed. For example, even to compute the value in the Black and Scholes market environment of foreign exchange European options with an arbitrary payoff or of foreign exchange American call and put options, a standard way to proceed is to use the corresponding binomial model. In this case, in order to compute the counterpart on the opposite side of the market of a previously computed foreign exchange option, whether or not one can employ a symmetry relationship and still maintain control of the precision is based solely on whether or not the symmetry relationship holds at all levels of the approximation. Fortunately, as we pointed out in Chapter 3, the symmetry relationships do hold at all levels of such approximation for approximation schemes based on the binomial model or on Bermudan options. Why is it that the control of the precision is based on the fact that the symmetry relationship holds at all levels of the approximation? Suppose hypothetically that this is not the case, that is, that the symmetry relationship holds only for the values of the options themselves and not for the corresponding members of their approximation schemes. Then the foreign exchange counterpart of a member of the approximation scheme of the option under consideration, found by applying the symmetry relationship, will not be a member of the corresponding approximation scheme for the counterpart of the option on the opposite side of the market.

The validity of the symmetry relationships at all levels of approximation becomes especially important when one does not have any a priori estimates for the precision of the chosen approximation scheme. In this case, one is forced to judge the precision of the chosen approximation scheme based on internal criteria, that is, on numerical experimentation that compares the numerical values of progressive members of this scheme.

The fact that the symmetry relationships are valid at all levels of approximation in the binomial model also allows for even further improvement of the computational efficiency of modern high-speed binomial methods for the valuation of various foreign exchange options. These methods achieve their high speed by dramatically reducing the number of steps needed in the binomial model while still preserving the desired precision. For one such method, see Price [31].

6.7   Reduced Cost of Software Development

The symmetry relationships also have applications to the development and testing of algorithms and software to value and analyze portfolios of options. The rapid change and high efficiency of modern financial markets, such as foreign exchange markets, presents developers of such software, both for in-house and commercial use, with the challenge of adapting their products rapidly to meet the time-constraint demands of users for the most up-to-date implementations. All this must be done while still maintaining high competitiveness in terms of the cost of the software development. In this regard, the symmetry relationships reduce significantly the cost of software development, not only by all the benefits described in this chapter, but also by cutting memory usage in half and significantly reducing coding and debugging time.

6.8   Guiding the Choice of Options with Improved Hedging Features

The symmetry relationships can be used to guide the choice in smoothing the corners of payoffs of European, Bermudan, and American options such as European, Bermudan, and American call and put options. The practical reason for this smoothing is that the resulting options with smoothed payoffs are easier to hedge for example in the Black and Scholes market environment. At the same time, these smoothed payoffs, and hence the values of these resulting options, are close to those of the original options with unsmoothed payoffs.

6.9   Validity of the Symmetry in any Financial Market

Finally, we note that the symmetry relationships are not limited to foreign exchange markets and, in fact, remain valid for any financial markets with exchange of arbitrary underlying securities.

Part II

Mathematical Matters

Chapter 7

Validity of the Symmetry Relationships for European Options

We begin the promised formal mathematical proofs of the results presented in Chapters 3 through 5. As we indicated in the Preface, we leave out some of the more subtle mathematical details, such as the exact domains and ranges of the evolution operators defining the market environments, partly because we do not wish to overload the book with too many fine details only of interest to a pure mathematician, but mainly because these technical details may obscure the structure of the symmetry, which is purely algebraic in nature.

In this chapter we prove the symmetry relationships for foreign exchange European options with general payoffs in the general market environment stated in Chapter 3. But first, since the case of European call and put options is of such paramount importance in practice and also for the sake of illustration, we give an independent proof of the symmetry relationships (1.1) and (1.2) for these options, even though they are particular instances of (3.5).

7.1 Direct Proof of the Symmetry Relationships for European Call and Put Options

We start with the first symmetry relationship in (1.1). Consider two portfolios of European options with inception time t and expiration time T. The first portfolio consists of a single European call option in currency d on currency f with strike price X, and the second portfolio consists of X units of European put option in currency f on currency d with strike price $1/X$.

Consider their values at the inception time t. The value of the first portfolio in currency d is $C_d(t, T, S, X)$ and the value of the second portfolio in currency f is $X\, P_f(t, T, S', 1/X)$, which in currency d is $SX\, P_f(t, T, 1/S, 1/X)$, where, by the no-arbitrage argument, the exchange rates S and S' at time t are assumed to be inverses to each other.

Now consider their values at expiration time T. The value of the first portfolio in currency d is $(S_T - X)^+$. The value of the second portfolio in currency f is $X(1/X - S_T')^+$, which in currency d is

$$S_T X \left(1/X - 1/S_T\right)^+ = (S_T - X)^+,$$

where again, by the no-arbitrage argument, the exchange rates S_T and S_T' at time T are assumed to be inverses to each other.

Thus the values of these portfolios in currency d are equal at expiration time T for all exchange rates $S_T > 0$. Therefore, by the no-arbitrage argument, their values

$$C_d(t, T, S, X) \quad \text{and} \quad SX\, P_f(t, T, 1/S, 1/X)$$

at inception time t in currency d must also be equal for all exchange rates $S > 0$, that is, the first symmetry relationship in (1.1) holds.

The remaining symmetry relationships in (1.1) and (1.2) for European call and put options can be proved analogously.

7.2   European Options: General Case

We now prove the symmetry relationships (3.5) for European options with general payoffs in the foreign exchange market. We prove only the first relationship since the proof of the second relationship is completely analogous.

Consider two portfolios of European options with inception time t and expiration time T. The first portfolio consists of a single European option in currency d on currency f with a payoff g_d. The second portfolio consists of a single European option in currency f on currency d with a payoff $g_f = \mathbf{K}g_d$. Their values in currency d at inception time t are $\mathcal{E}_d(t, T, g_d)$ and $\mathbf{K}\,\mathcal{E}_f(t, T, \mathbf{K}g_d)$.

Now consider their values at expiration. The value of the first portfolio in currency d is g_d. The value of the second portfolio in currency f is $g_f = \mathbf{K}g_d$, and so in currency d it is

$$\mathbf{K}(\mathbf{K}g_d) = \mathbf{K}^2 g_d = g_d,$$

where we have used the fact that $\mathbf{K}^2$ is the identity operator on Π. Thus the values of these portfolios in currency d are equal at expiration time T. Therefore, by the no-arbitrage argument, their values in currency d at inception time t must also be equal. This proves the first relationship in (3.5).

7.3   European Call and Put Options

The symmetry relationships (1.1) and (1.2) for European call and put options in the foreign exchange market are particular cases of the symmetry relationships (3.5) for European options with general payoffs. In order to see this, we first note that the linearity of the evolution operators $\mathbf{V}_d(t,T)$ and $\mathbf{V}_f(t,T)$ implies that the values $\mathcal{E}_d(t,T,g_d)$ and $\mathcal{E}_f(t,T,g_f)$ of European options are homogeneous of degree one as functions of their payoffs g_d and g_f, that is,

$$(7.1) \quad \begin{aligned} \mathcal{E}_d(t,T,\lambda g_d) &= \lambda\,\mathcal{E}_d(t,T,g_d), \\ \mathcal{E}_f(t,T,\lambda g_f) &= \lambda\,\mathcal{E}_f(t,T,g_f), \end{aligned}$$

for all nonnegative λ. We recall that the linearity of the evolution operators $\mathbf{V}_d(t,T)$ and $\mathbf{V}_f(t,T)$ itself was, in turn, established by the no-arbitrage argument for European options.

We will prove only the first symmetry relationship in (1.1) since the proofs of the second symmetry relationship in (1.1) and the symmetry relationships in (1.2) are completely analogous. The proof follows from the following chain of equalities:

$$\begin{aligned} C_d(t,T,S,X) &= \mathcal{E}_d(t,T,(\cdot - X)^+)(S) \\ &= \mathbf{K}\,\mathcal{E}_f(t,T,\mathbf{K}(\cdot - X)^+)(S) \\ &= \mathbf{K}\,\mathcal{E}_f(t,T,X(1/X - \cdot)^+)(S) \\ &= X\,\mathbf{K}\,\mathcal{E}_f(t,T,(1/X - \cdot)^+)(S) \\ &= SX\,P_f(t,T,1/S,1/X), \end{aligned}$$

where the first and last equalities are due to the definitions of European call and put options, the second equality is due to the first symmetry relationship in (3.5), the third equality is due to the first symmetry relationship in (3.2), and the fourth equality is due to the second homogeneity relationship in (7.1) and the linearity of the Kelvin transform.

Finally we note, that the symmetry relationships in (3.4) follow from the symmetry relationships in (3.5) combined with the definitions of the operators $\mathbf{V}_d(t,T)$ and $\mathbf{V}_f(t,T)$ as the unique linear extensions to a linear subspace of Π of the operators that map g_d and g_f in Π_+ to $\mathcal{E}_d(t,T,g_d)$ and $\mathcal{E}_f(t,T,g_f)$.

7.4 Symmetry in the Binomial Market Environment

Now we show that the binomial market environments $\mathbf{V}_d^B$ and $\mathbf{V}_f^B$ provide a consistent model of a foreign exchange option market, that is, we show that the symmetry relationships (3.6) and (3.7) are valid.

We first prove the symmetry relationships in (3.6). It is enough to prove them only for neighboring times t and T from the trading time set $\mathcal{T}$, that is, it is enough to prove only the symmetry relationships

$$
\begin{aligned}
\mathbf{V}_d^B(t_l, t_{l+1}) &= \mathbf{K}\mathbf{V}_f^B(t_l, t_{l+1})\mathbf{K}, \\
\mathbf{V}_f^B(t_l, t_{l+1}) &= \mathbf{K}\mathbf{V}_d^B(t_l, t_{l+1})\mathbf{K}.
\end{aligned}
$$
(7.2)

The reason for this is that, due to the fact that $\mathbf{K}^2$ is the identity operator on Π and due to the definitions of the operators $\mathbf{V}_d^B(t, T)$ and $\mathbf{V}_f^B(t, T)$ in (2.5), these operators can be represented as follows:

(7.3)
$$
\begin{aligned}
\mathbf{V}_d^B(t, T) =& \\
&\mathbf{K}^2\mathbf{V}_d^B(t = t_i, t_{i+1})\mathbf{K}^2\mathbf{V}_d^B(t_{i+1}, t_{i+2})\mathbf{K}^2 \ldots \mathbf{K}^2\mathbf{V}_d^B(t_{j-1}, t_j = T)\mathbf{K}^2, \\
\mathbf{V}_f^B(t, T) =& \\
&\mathbf{K}^2\mathbf{V}_f^B(t = t_i, t_{i+1})\mathbf{K}^2\mathbf{V}_f^B(t_{i+1}, t_{i+2})\mathbf{K}^2 \ldots \mathbf{K}^2\mathbf{V}_f^B(t_{j-1}, t_j = T)\mathbf{K}^2.
\end{aligned}
$$

Substituting (7.2) into (7.3), we obtain

$$
\begin{aligned}
\mathbf{V}_d^B(t, T) =& \\
&\mathbf{K}\mathbf{V}_f^B(t = t_i, t_{i+1})\mathbf{V}_f^B(t_{i+1}, t_{i+2}) \ldots \mathbf{V}_f^B(t_{j-1}, t_j = T)\mathbf{K} = \mathbf{K}\mathbf{V}_f^B(t, T)\mathbf{K}, \\
\mathbf{V}_f^B(t, T) =& \\
&\mathbf{K}\mathbf{V}_d^B(t = t_i, t_{i+1})\mathbf{V}_d^B(t_{i+1}, t_{i+2}) \ldots \mathbf{V}_d^B(t_{j-1}, t_j = T)\mathbf{K} = \mathbf{K}\mathbf{V}_d^B(t, T)\mathbf{K},
\end{aligned}
$$

by definition of the operators $\mathbf{V}_d^B(t, T)$ and $\mathbf{V}_f^B(t, T)$ in (2.5). That is, (3.6) holds.

Returning to the proofs of the symmetry relationships in (7.2), we prove only the first one since the proof of the second is completely analogous. According to the definitions of $\mathbf{V}_d^B(t_l, t_{l+1})$ and $\mathbf{V}_f^B(t_l, t_{l+1})$ in (2.6) and the definition of the Kelvin transform $\mathbf{K}$, the first relationship in (7.2) can be rewritten as follows:

$$(\mathbf{V}_d^B(t_l, t_{l+1})h_d)(S) =$$
$$\frac{p'_{u,l}(1/S)}{\rho'_{u,l}(1/S)}u'_l(1/S)h_d\left(\frac{1}{u'_l(1/S)}S\right) + \frac{p'_{v,l}(1/S)}{\rho'_{v,l}(1/S)}v'_l(1/S)h_d\left(\frac{1}{v'_l(1/S)}S\right).$$

To prove this equality and hence prove the first relationship in (7.2), we note that, by virtue of the definition of $\mathbf{V}_d^B(t_l, t_{l+1})$ in (2.6) and the definitions of the functions u'_l and v'_l, it is enough to prove the identities

$$\frac{p'_{u,l}(1/S)}{\rho'_{u,l}(1/S)}u'_l(1/S) = \frac{p_{u,l}(S)}{\rho_{u,l}(S)},$$

$$\frac{p'_{v,l}(1/S)}{\rho'_{v,l}(1/S)}v'_l(1/S) = \frac{p_{v,l}(S)}{\rho_{v,l}(S)}.$$

These identities follow easily from the definitions of the risk-neutral transition probabilities $p_{u,l}$, $p_{v,l}$, $p'_{u,l}$, and $p'_{v,l}$ in (2.3) and (2.4). This completes the proof of symmetry relationships in (3.6).

The symmetry relationships in (3.7) follow directly from symmetry relationships in (3.6) using the definitions of the operators $\hat{\mathbf{V}}_d^B(t,T)$ and $\hat{\mathbf{V}}_f^B(t,T)$ in (2.17) and of $\hat{\mathbf{K}}_{f\to d}(t)$ and $\hat{\mathbf{K}}_{d\to f}(t)$ in (3.8).

7.5   Symmetry in the Black and Scholes Market Environment

In this Section we show that the Black and Scholes market environments $\mathbf{V}_d^{BS}$ and $\mathbf{V}_f^{BS}$ provide a consistent model of a foreign exchange option market, that is, we show that the symmetry relationships in (3.10) are valid. We prove only the first symmetry relationship in (3.10) since the proof of the second symmetry relationship is completely analogous.

According to the definitions of the evolution operators $\mathbf{V}_d^{BS}(t,T)$ and $V_f^{BS}(t,T)$ in (2.20) and (2.21), what we need to prove is that

$$\frac{e^{-r_d\tau}}{\sigma\sqrt{2\pi\tau}}\int_0^\infty x^{-1}e^{-(\log(x/S)-(r_d-r_f-\sigma^2/2)\tau)^2/2\sigma^2\tau}h_d(x)\,dx$$

$$= S\frac{e^{-r_f\tau}}{\sigma\sqrt{2\pi\tau}}\int_0^\infty x^{-1}e^{-(\log(x/S^{-1})-(r_f-r_d-\sigma^2/2)\tau)^2/2\sigma^2\tau}(x\,h_d(1/x))\,dx$$

for any admissible h_d in Π. This is achieved by making the substitution $x \mapsto 1/x$ in the second integral.

Finally, we comment that as we indicated in Section 3.3, the symmetry relationships (3.10) can be reformulated as the symmetry relationships (3.18). To prove the symmetry relationships (3.18) we have to show that the equalities

$$L_d^{BS} h_d = \mathbf{K} L_f^{BS} \mathbf{K} h_d$$
$$L_f^{BS} h_f = \mathbf{K} L_d^{BS} \mathbf{K} h_f,$$

hold for any, for example, sufficiently smooth, functions $h_d = h_d(S)$ and $h_f(S')$ in Π, where the operators L_d^{BS} and L_f^{BS} were defined in Section 3.3 as

$$L_d^{BS} = \frac{1}{2} \sigma^2 S^2 \frac{\partial^2}{\partial S^2} + (r_d - r_f) S \frac{\partial}{\partial S} - r_d,$$

and

$$L_f^{BS} = \frac{1}{2} \sigma^2 S'^2 \frac{\partial^2}{\partial S'^2} + (r_f - r_d) S' \frac{\partial}{\partial S'} - r_f.$$

In the case of sufficiently smooth functions, this, in turn, can be shown with the help of the definition of the Kelvin transform $\mathbf{K}$ by straightforward differentiation.

This completes the proofs of the symmetry relationships for European options with general payoffs in a foreign exchange market.

Chapter 8

Validity of the Symmetry Relationships for Bermudan and American Options

In this chapter we prove the symmetry relationships presented in Chapter 3 for foreign exchange Bermudan and American options with general time-dependent payoffs in general market environments $\mathbf{V}_d$ and $\mathbf{V}_f$ satisfying the intervention condition.

8.1 Bermudan Options: General Case

In this section we prove the symmetry relationships in (3.19) for foreign exchange Bermudan options with general time-dependent payoffs in general market environments $\mathbf{V}_d$ and $\mathbf{V}_f$ satisfying the intervention condition in Chapter 3.

We prove only the first relationship in (3.19) since the proof of the second relationship is completely analogous. In order to do this, we need further preparation.

Firstly, we need the explicit expression for the value of a Bermudan option in a general market environment satisfying the intervention condition. This explicit expression was obtained by Kholodnyi in [18], and in the foreign exchange setting it reads:

(8.1)
$$\mathcal{B}_d(t,T,E,g_d) =$$
$$\mathbf{V}_d(t,t_0)\mathbf{M}_{g_d^{(0)}}\mathbf{V}_d(t_0,t_1)\mathbf{M}_{g_d^{(1)}}\mathbf{V}_d(t_1,t_2)\ldots\mathbf{M}_{g_d^{(n-1)}}\mathbf{V}_d(t_{n-1},T)g_d^{(n)},$$
$$\mathcal{B}_f(t,T,E,g_f) =$$
$$\mathbf{V}_f(t,t_0)\mathbf{M}_{g_f^{(0)}}\mathbf{V}_f(t_0,t_1)\mathbf{M}_{g_f^{(1)}}\mathbf{V}_f(t_1,t_2)\ldots\mathbf{M}_{g_f^{(n-1)}}\mathbf{V}_f(t_{n-1},T)g_f^{(n)},$$

where the nonlinear operator $\mathbf{M}_h : \Pi \to \Pi$, with h in Π, is defined by

$$(\mathbf{M}_h g)(S) = \max\{h(S), g(S)\}, \quad g \in \Pi,\ S > 0$$

and where $g_d^{(i)} = g_{d,t_i}$ and $g_f^{(i)} = g_{f,t_i}$ for $i = 0,1,\ldots,n$ with each exercise time t_i contained in the exercise time set $E = \{t_i : i = 0,1,\ldots,n\}$.

Note that the operator $\mathbf{M}_h$ for any h in Π preserves the nonnegative cone Π_+ of Π.

The above expressions for the values of the foreign exchange Bermudan options have the following financial interpretation described in [18]. For the domestic side of the market, as an example, at each exercise time t_{i-1} in the exercise time set E, an investor has two choices: either

- to exercise the option and receive the payoff $g_{d,t_{i-1}}(S_{t_{i-1}})$ where the exchange rate between currency d and currency f is $S_{t_{i-1}}$, or

- to keep the option until the next exercise time t_i as a European option in currency d on currency f with a payoff being the value of the Bermudan option at exercise time t_i as a function of S_{t_i}.

The maximum of the values of this European option at its inception time t_{i-1} and the payoff $g_{d,t_{i-1}}(S_{t_{i-1}})$ is therefore the value of the Bermudan option at the exercise time t_{i-1} as a function of $S_{t_{i-1}}$.

Note that the values of the Bermudan options in (8.1) are represented in the form of a universal contingent claim as a multiplicative measure given by the product integral of, generally speaking, nonlinear and noncommutative operators. As we pointed earlier in the book, for the definition and detailed discussion of a universal contingent claim and of multiplicative measure see [22].

We also note that the value $\mathcal{B}(t,T,E,g)$ of the abstract Bermudan option with inception time t, expiration time T, and with a (time-dependent) abstract payoff $g : E \to \Pi_+$, as a Bermudan option on a vector lattice of all generalized payoffs, is given explicitly in [18] by

$$\mathcal{B}(t, T, E, g) =$$
$$\mathcal{V}(t, t_0)\mathbf{M}_{g^{(0)}}\mathcal{V}(t_0, t_1)\mathbf{M}_{g^{(1)}}\mathcal{V}(t_1, t_2)\ldots\mathbf{M}_{g^{(n-1)}}\mathcal{V}(t_{n-1}, T)g^{(n)},$$

where $g^{(i)} = g_{t_i}$ for $i = 0, 1, \ldots, n$, and where the nonlinear operator $\mathbf{M}_h$: $\Pi \to \Pi$, with h in Π, is defined by

$$\mathbf{M}_h g = h \vee g, \quad g \in \Pi.$$

Here $\vee$ denotes the lattice operation of supremum in the vector lattice of all abstract payoffs Π. We comment that the abstract evolution operators $\mathcal{V}(t, T)$ in the above expression were assumed to satisfy the intervention condition.

Secondly, we need the following identity:

(8.2) $$\mathbf{K}_c \mathbf{M}_h\, g = \mathbf{M}_{\mathbf{K}_c h}\, \mathbf{K}_c g, \quad c > 0,$$

for any payoffs h and g in Π.

This identity for $c = 1$ has the financial interpretation that if we have the choice between two payoffs in Π on either side of the foreign exchange market, then we would choose the maximum of the two. Then the identity states that the value of this maximum payoff converted to the opposite side of the market is the maximum of the two payoffs separately converted to the opposite side.

Mathematically the identity (8.2) states that the Kelvin transform $\mathbf{K}_c$ with arbitrary $c > 0$, as a faithful representation of the cyclic group $\mathbb{Z}_2$ in the linear space of all isomorphisms on Π as a vector space, and, as a linear operator that preserves the nonnegative cone Π_+ of Π, is an isomorphism of Π as the vector lattice defined in Chapter 3.

The proof of the identity (8.2) follows from the chain of equalities:

$$(\mathbf{K}_c \mathbf{M}_h f)(S) = (S/\sqrt{c})\max(h(c/S), f(c/S))$$
$$= \max((S/\sqrt{c})h(c/S), (S/\sqrt{c})f(c/S))$$
$$= (\mathbf{M}_{\mathbf{K}_c h}\, \mathbf{K}_c f)(S),$$

where the first and the last equalities follow from the definition of the Kelvin transform $\mathbf{K}_c$ and the operator $\mathbf{M}_h$, and the second equality follows from the definition of the maximum function.

With this preparation, the following chain of equalities completes the proof of the first symmetry relationship in (3.19) for foreign exchange Bermudan options:

$$\mathcal{B}_d(t, T, E, g_d)$$
$$= \mathbf{V}_d(t, t_0)\mathbf{M}_{g_d^{(0)}}\mathbf{V}_d(t_0, t_1)\mathbf{M}_{g_d^{(1)}}\mathbf{V}_d(t_1, t_2)\dots$$
$$\dots\mathbf{M}_{g_d^{(n-1)}}\mathbf{V}_d(t_{n-1}, T)g_d^{(n)},$$
$$= \mathbf{K}^2\mathbf{V}_d(t, t_0)\mathbf{K}^2\mathbf{M}_{g_d^{(0)}}\mathbf{V}_d(t_0, t_1)\mathbf{K}^2\mathbf{M}_{g_d^{(1)}}\mathbf{V}_d(t_1, t_2)\mathbf{K}^2\dots$$
$$\dots\mathbf{K}^2\mathbf{M}_{g_d^{(n-1)}}\mathbf{V}_d(t_{n-1}, T)\mathbf{K}^2 g_d^{(n)},$$
$$= \mathbf{K}\mathbf{K}\mathbf{V}_d(t, t_0)\mathbf{K}\mathbf{M}_{\mathbf{K}g_d^{(0)}}\mathbf{K}\mathbf{V}_d(t_0, t_1)\mathbf{K}\mathbf{M}_{\mathbf{K}g_d^{(1)}}\mathbf{K}\mathbf{V}_d(t_1, t_2)\mathbf{K}\dots$$
$$\dots\mathbf{K}\mathbf{M}_{\mathbf{K}g_d^{(n-1)}}\mathbf{K}\mathbf{V}_d(t_{n-1}, T)\mathbf{K}\mathbf{K}g_d^{(n)},$$
$$= \mathbf{K}\mathbf{V}_f(t, t_0)\mathbf{M}_{\mathbf{K}g_d^{(0)}}\mathbf{V}_f(t_0, t_1)\mathbf{M}_{\mathbf{K}g_d^{(1)}}\mathbf{V}_f(t_1, t_2)\dots$$
$$\dots\mathbf{M}_{\mathbf{K}g_d^{(n-1)}}\mathbf{V}_f(t_{n-1}, T)\mathbf{K}g_d^{(n)}$$
$$= \mathbf{K}\mathcal{B}_f(t, T, E, \mathbf{K}g_d),$$

where the first and the last equalities are due to the relationships in (8.1), the second equality is due to the fact that $\mathbf{K}^2$ is the identity operator on Π, the third equality is due to the identity (8.2), and the fourth equality is due to the second symmetry relationship in (3.4).

It is clear from the preceding proof, that with the help of the symmetry relationships in (3.6) and (3.10) the symmetry relationships (3.19) hold for foreign exchange Bermudan options valued in the two major cases of market environments, the binomial market environments $\mathbf{V}_d^B$ and $\mathbf{V}_f^B$ and the Black and Scholes market environments $\mathbf{V}_d^{BS}$ and $\mathbf{V}_f^{BS}$.

Using the terminology of category theory, the preceding proof of the first symmetry relationship in (3.19) is illustrated by the commutative or no-arbitrage diagram of financial operations in Figure 8.1. A similar commutative or no-arbitrage diagram of financial operations can illustrate the proof of the second symmetry relationship in (3.19).

8.2   Bermudan Call and Put Options

In this section we prove the fact that the symmetry relationships in (3.19) take the form of relationships (1.1) and (1.2) for the particular cases of Bermudan call and put options.

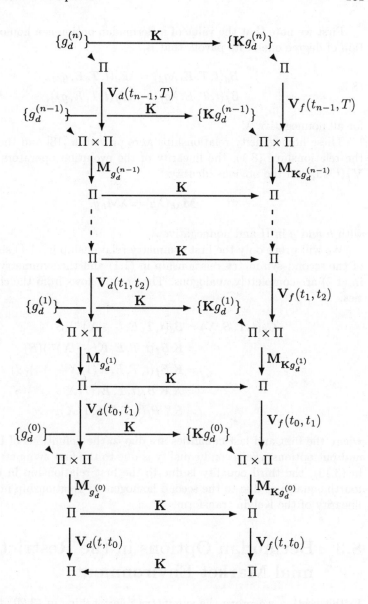

Figure 8.1: **Commutative or no-arbitrage diagram for the symmetry for Bermudan options.**

First we note that the value of a Bermudan option is a homogeneous function of degree one of its payoff, that is,

$$
\begin{aligned}
\mathcal{B}_d(t, T, E, \lambda g_d) &= \lambda \mathcal{B}_d(t, T, E, g_d) \\
\mathcal{B}_f(t, T, E, \lambda g_f) &= \lambda \mathcal{B}_f(t, T, E, g_f),
\end{aligned}
$$

(8.3)

for all nonnegative λ.

These homogeneity relationships were proved in [18] and they follow from the relationships (8.1), the linearity of the evolution operators $\mathbf{V}_d(t, T)$ and $\mathbf{V}_f(t, T)$, and the obvious identity

$$
\mathbf{M}_{\lambda h}(\lambda g) = \lambda \mathbf{M}_h g,
$$

with h and g in Π and nonnegative λ.

We will prove only the first symmetry relationship in (1.1) since the proofs of the second symmetry relationship in (1.1) and the symmetry relationships in (1.2) are completely analogous. The proof follows from the chain of equalities:

$$
\begin{aligned}
C_d(t, T, S, X) &= \mathcal{B}_d(t, T, E, (\cdot - X)^+)(S) \\
&= \mathbf{K}\, \mathcal{B}_f(t, T, E, \mathbf{K}(\cdot - X)^+)(S) \\
&= \mathbf{K}\, \mathcal{B}_f(t, T, E, X(1/X - \cdot)^+)(S) \\
&= X\, \mathbf{K}\, \mathcal{B}_f(t, T, E, (1/X - \cdot)^+)(S) \\
&= SX\, P_f(t, T, 1/S, 1/X),
\end{aligned}
$$

where the first and last equalities are due to the definitions of Bermudan call and put options, the second equality is due to the first symmetry relationship in (3.19), the third equality is due to the first relationship in (3.2), and the fourth equality is due to the second homogeneity relationship in (8.3) and the linearity of the Kelvin transform.

8.3 Bermudan Options in the Restricted Binomial Market Environment

In this section we prove the symmetry relationships in (3.20) for foreign exchange Bermudan options in the binomial market environments $\mathbf{V}_d^B$ and $\mathbf{V}_f^B$ restricted to the binomial trees Γ_d and Γ_f, that is, in the restricted binomial market environments $\hat{\mathbf{V}}_d^B$ and $\hat{\mathbf{V}}_f^B$.

First we note that, as it was shown in [18],

(8.4)
$$\hat{\mathcal{B}}_d(t, T, E, \hat{g}_d) =$$
$$\hat{\mathbf{V}}_d^B(t, t_0)\hat{\mathbf{M}}_{\hat{g}_d^{(0)}}\hat{\mathbf{V}}_d^B(t_0, t_1)\hat{\mathbf{M}}_{\hat{g}_d^{(1)}}\hat{\mathbf{V}}_d^B(t_1, t_2)\ldots\hat{\mathbf{M}}_{\hat{g}^{(n-1)}}\hat{\mathbf{V}}_d^B(t_{n-1}, T)\hat{g}_d^{(n)},$$

$$\hat{\mathcal{B}}_f(t, T, E, \hat{g}_f) =$$
$$\hat{\mathbf{V}}_f^B(t, t_0)\hat{\mathbf{M}}_{\hat{g}_f^{(0)}}\hat{\mathbf{V}}_f^B(t_0, t_1)\hat{\mathbf{M}}_{\hat{g}_f^{(1)}}\hat{\mathbf{V}}_f^B(t_1, t_2)\ldots\hat{\mathbf{M}}_{\hat{g}^{(n-1)}}\hat{\mathbf{V}}_f^B(t_{n-1}, T)\hat{g}_f^{(n)},$$

where $\hat{g}_d^{(i)} = Q_d(t_i)g_d^{(i)}$ and $\hat{g}_f^{(i)} = Q_f(t_i)g_f^{(i)}$ with $g_d^{(i)} = g_{d,t_i}$ and $g_f^{(i)} = g_{f,t_i}$, and where the operators

$$\hat{\mathbf{M}}_{\hat{g}_d^{(i)}} : \Pi_d(t_i) \to \Pi_d(t_i) \quad \text{and} \quad \hat{\mathbf{M}}_{\hat{g}_f^{(i)}} : \Pi_f(t_i) \to \Pi_f(t_i),$$

called the restrictions of the operators $\mathbf{M}_{g_d^{(i)}}$ and $\mathbf{M}_{g_f^{(i)}}$ to $\Pi_d(t_i)$ and $\Pi_f(t_i)$, respectively, are defined by

$$\hat{\mathbf{M}}_{\hat{g}_d^{(i)}} Q_d(t_i) = Q_d(t_i)\,\mathbf{M}_{g_d^{(i)}},$$
$$\hat{\mathbf{M}}_{\hat{g}_f^{(i)}} Q_f(t_i) = Q_f(t_i)\,\mathbf{M}_{g_f^{(i)}}.$$

Then the symmetry relationships in (3.20) follow directly from the symmetry relationships in (3.19), the relationships (8.4), and the definitions of the operators $\hat{\mathbf{V}}_d^B(t, T)$ and $\hat{\mathbf{V}}_f^B(t, T)$ in (2.17) and of $\hat{\mathbf{K}}_{f\to d}(t)$ and $\hat{\mathbf{K}}_{d\to f}(t)$ in (3.8).

The proof of the symmetry relationships (3.22) and (3.23) for Bermudan call and put options in the binomial market environments $\mathbf{V}_d^B$ and $\mathbf{V}_f^B$ restricted to the binomial trees Γ_d and Γ_f, that is, in the restricted binomial market environments $\hat{\mathbf{V}}_d^B$ and $\hat{\mathbf{V}}_f^B$, follows directly from the symmetry relationships (1.1) and (1.2) for Bermudan call and put options in the binomial market environments $\mathbf{V}_d^B$ and $\mathbf{V}_f^B$ and the expressions (3.21) for the values of Bermudan options in the restricted binomial market environments $\hat{\mathbf{V}}_d^B$ and $\hat{\mathbf{V}}_f^B$.

8.4   American Options: General Case

In this section we prove the symmetry relationships in (3.24) for foreign exchange American options with general time-dependent payoffs in general market environments $\mathbf{V}_d$ and $\mathbf{V}_f$ satisfying the intervention condition (3.12).

The proof is based on the fact indicated in [18] that American options can be financially viewed as a certain limit of the corresponding Bermudan options when their exercise time sets E, roughly speaking, fill all the exercise time interval $[t, T]$:

$$
\begin{aligned}
A_d(t, T, g_d) &= \lim B_d(t, T, E, \hat{g}_d), \\
A_f(t, T, g_f) &= \lim B_f(t, T, E, \hat{g}_f),
\end{aligned}
$$

(8.5)

where $\hat{g}_d$, $\hat{g}_f : E \to \Pi_+$ are the restrictions of g_d, $g_f : [t, T] \to \Pi_+$ to the exercise time set E. For details of mathematical nature of these limits see [18] and [14].

To finish the proof, it is enough to notice that, since the symmetry relationships in (3.19) hold for any members $B_d(t, T, E, \hat{g}_d)$ and $B_f(t, T, E, \hat{g}_f)$ with $\hat{g}_d$ and $\hat{g}_f$ related via the Kelvin transform, then the symmetry relationships in (3.24) holds for their limits $A_d(t, T, g_d)$ and $A_f(t, T, g_f)$ with g_d and g_f related via the Kelvin transform.

It is clear from the preceding proof, that with the help of the symmetry relationships (3.10) the symmetry relationships (3.24) hold for foreign exchange American options valued in one of the major cases of market environments, the Black and Scholes market environments $\mathbf{V}_d^{BS}$ and $\mathbf{V}_f^{BS}$. We note again that American options do not exist in the other major case of the market environment, the binomial market environment, since these options require the trading time sets to contain an entire time interval but the trading time sets in the binomial market environment are discrete.

Note that the values of the American options in (8.5), taking into account the expressions (8.1) for the values of Bermudan options, are represented in the form of a universal contingent claim as a multiplicative measure given by product integral of, generally speaking, nonlinear and noncommutative operators. Once again, for the definition and detailed discussion of a universal contingent claim and of multiplicative measure see [22].

We also note that the value $A(t, T, E, g)$ of an abstract American option with inception time t, expiration time T, and with a (time-dependent) abstract payoff $g : E \to \Pi_+$, as an American option on a vector lattice of all generalized payoffs is given explicitly in [18]. It is based on the fact that an abstract American option can be viewed as a certain limit of the corresponding Bermudan options when their exercise time sets E, roughly speaking, fill all the exercise time interval $[t, T]$:

(8.6) $A(t, T, g) = \lim B(t, T, E, \hat{g})$

where here $\hat{g} : E \to \Pi_+$ are the restrictions to the exercise time set E of

$g : [t, T] \to \Pi_+$. In order to have a meaningful concept of a limit, it is necessary to assume that Π is a topological vector lattice.

8.5 American Call and Put Options

In this section we prove the fact that the symmetry relationships in (3.24) take the form of equalities (1.1) and (1.2) for the particular cases of American call and put options.

First we note that the value of an American option is a homogeneous function of degree one of its payoff, that is,

(8.7)
$$\mathcal{A}_d(t, T, \lambda g_d) = \lambda \, \mathcal{A}_d(t, T, g_d),$$
$$\mathcal{A}_f(t, T, \lambda g_f) = \lambda \, \mathcal{A}_f(t, T, g_f),$$

for all nonnegative λ. These homogeneity relationships were proved in [18] and they follow from the homogeneity relationships for Bermudan options in (8.3) combined with the limit relationships in (8.5).

We shall prove only the first symmetry relationship in (1.1) since the proofs of the second symmetry relationship in (1.1) and the symmetry relationships in (1.2) are completely analogous. The proof follows from the chain of equalities:

$$\begin{aligned}
C_d(t, T, S, X) &= \mathcal{A}_d(t, T, (\cdot - X)^+)(S) \\
&= \mathbf{K}\, \mathcal{A}_f(t, T, \mathbf{K}(\cdot - X)^+)(S) \\
&= \mathbf{K}\, \mathcal{A}_f(t, T, X(1/X - \cdot)^+)(S) \\
&= X\, \mathbf{K}\, \mathcal{A}_f(t, T, (1/X - \cdot)^+)(S) \\
&= SX\, P_f(t, T, 1/S, 1/X),
\end{aligned}$$

where the first and last equalities are due to the definitions of American call and put options, the second equality is due to the first symmetry relationship in (3.24), the third equality is due to the first relationship in (3.2), and the fourth equality is due to the second homogeneity relationship in (8.7) and the linearity of the Kelvin transform.

Note that the symmetry relationships (1.1) and (1.2) for American call and put options also follow from the symmetry relationships (1.1) and (1.2) for Bermudan call and put options and the limit relationships in (8.5).

8.6   The Semilinear Evolution Equation for American Options

Although the symmetry relationships for foreign exchange American options in the Black and Scholes market environment are a particular case of the symmetry relationships (3.24) and symmetry relationships (3.4) as given by (3.10) and as such were proved in Section 8.4, we derive them in this Section, as an illustration, directly from the semilinear evolution equation for such American options. This semilinear evolution equation for American options with general time-dependent payoffs in the Black and Scholes market environment, as well as in a general market environment, was introduced and studied by Kholodnyi in [17], [20] and [12]. In the Black and Scholes market environment and in the foreign exchange setting this equation takes the form:

(8.8)

$$\frac{\partial}{\partial t}V_d + \frac{1}{2}\sigma^2 S^2 \frac{\partial^2}{\partial S^2}V_d + (r_d - r_f)S\frac{\partial}{\partial S}V_d - r_d V_d$$
$$+ \rho_d(S,t)\,H(g_d(S,t) - V_d) = 0, \quad S > 0,\ t \in [0,T),$$
$$V_d(S,T) = g_d(S,T),$$

and

(8.9)

$$\frac{\partial}{\partial t}V_f + \frac{1}{2}\sigma^2 S'^2 \frac{\partial^2}{\partial S'^2}V_f + (r_f - r_d)S'\frac{\partial}{\partial S'}V_f - r_f V_f$$
$$+ \rho_f(S',t)\,H(g_f(S',t) - V_f) = 0, \quad S' > 0,\ t \in [0,T),$$
$$V_f(S',t) = g_f(S',T),$$

where H is the Heaviside function, and where

$$\rho_d(S,t) = (-(\frac{\partial}{\partial t}g_d + \frac{1}{2}\sigma^2 S^2 \frac{\partial^2}{\partial S^2}g_d + (r_d - r_f)S\frac{\partial}{\partial S}g_d - r_d g_d)(S,t))^+$$

and

$$\rho_f(S',t) = (-(\frac{\partial}{\partial t}g_f + \frac{1}{2}\sigma^2 S'^2 \frac{\partial^2}{\partial S'^2}g_f + (r_f - r_d)S'\frac{\partial}{\partial S'}g_f - r_f g_f)(S',t))^+.$$

The semilinear evolution equation (8.8), or more precisely, the Cauchy problem for the semilinear evolution equation (8.8), determines the value in the Black and Scholes market environment $\mathbf{V}_d^{BS}$ of the American option in currency d on currency f with inception time t, expiration time T, and with the general time-dependent payoff g_d. Similarly, the semilinear evolution equation (8.9), or more precisely, the Cauchy problem for the semilinear evolution equation (8.9), determines the value in the Black and Scholes market environment $\mathbf{V}_f^{BS}$ of the American option in currency f on currency d with inception time t, expiration time T, and with the general time-dependent payoff g_f.

Note that in general market environments $\mathbf{V}_d$ and $\mathbf{V}_f$ generated by the families of linear operators

$$\mathbf{L}_d = \{\mathbf{L}_d(t) : \Pi \to \Pi \,|\, t \in \mathcal{T}\}$$

and

$$\mathbf{L}_f = \{\mathbf{L}_f(t) : \Pi \to \Pi \,|\, t \in \mathcal{T}\},$$

the generators

$$L_d^{BS} = \frac{1}{2}\sigma^2 S^2 \frac{\partial^2}{\partial S^2} + (r_d - r_f)S\frac{\partial}{\partial S} - r_d$$

and

$$L_f^{BS} = \frac{1}{2}\sigma^2 S'^2 \frac{\partial^2}{\partial S'^2} + (r_f - r_d)S'\frac{\partial}{\partial S'} - r_f.$$

of the Black and Scholes market environments $\mathbf{V}_d^{BS}$ and $\mathbf{V}_f^{BS}$ in the semilinear evolution equations (8.8) and (8.9) are replaced by the generators $\mathbf{L}_d(t)$ and $\mathbf{L}_f(t)$ (see [17], [20] and [12]).

Note also that in view of the previous paragraph we have so far two seemingly different descriptions of American options in general market environments $\mathbf{V}_d$ and $\mathbf{V}_f$ generated by the families of linear operators $\mathbf{L}_d$ and $\mathbf{L}_f$. One is in terms of the semilinear evolution equation for American options, and another is in terms of the limit of Bermudan options in (8.5). As a relevant example we have used American options in the foreign exchange setting. The equivalency of these two descriptions was established in [18] and [14].

We comment that the general time-dependent payoffs g_d and g_d in the semilinear evolution equations (8.8) and (8.9) should be in the class of admissible payoffs. The admissible payoffs can be characterized, for example, by their regularity in the exchange rate and time variables and by their growth at zero and at infinity in the exchange rate variables. The analysis of this admissibility issue is presented in [17], [20] and [12].

We also comment that the nonlinear terms

$$\rho_d(S,t)\,H(g_d(S,t) - V_d) \quad \text{and} \quad \rho_f(S',t)\,H(g_f(S',t) - V_f)$$

in the semilinear evolution equations (8.8) and (8.9) have the financial meaning of the cash flows that should be received to compensate for the losses due to the holding of the American options with the payoffs g_d and g_f unexercised when it is optimal to exercise these options, that is, to exercise the right to receive the payoffs g_d and g_f (see [17]).

In the particular cases of the American call and put options the semilinear evolution equations (8.8) and (8.9) take the form (see [17]):

(8.10)
$$\frac{\partial}{\partial t}V_d + \frac{1}{2}\sigma^2 S^2 \frac{\partial^2}{\partial S^2}V_d + (r_d - r_f)S\frac{\partial}{\partial S}V_d - r_d V_d$$
$$+ (r_f S - r_d X)^+ H((S - X)^+ - V_d) = 0, \quad S > 0, \; t \in [0,T),$$
$$V_d(S,T) = (S - X)^+,$$

(8.11)
$$\frac{\partial}{\partial t}V_d + \frac{1}{2}\sigma^2 S^2 \frac{\partial^2}{\partial S^2}V_d + (r_d - r_f)S\frac{\partial}{\partial S}V_d - r_d V_d$$
$$+ (r_d X - r_f S)^+ H((X - S)^+ - V_d) = 0, \quad S > 0, \; t \in [0,T),$$
$$V_d(S,T) = (X - S)^+,$$

and

(8.12)
$$\frac{\partial}{\partial t}V_f + \frac{1}{2}\sigma^2 S'^2 \frac{\partial^2}{\partial S'^2}V_f + (r_f - r_d)S'\frac{\partial}{\partial S'}V_f - r_f V_f$$
$$+ (r_d S' - r_f X') H((S' - X')^+ - V_f) = 0, \quad S' > 0, \; t \in [0,T),$$
$$V_f(S',T) = (S' - X')^+,$$

(8.13)
$$\frac{\partial}{\partial t}V_f + \frac{1}{2}\sigma^2 S'^2 \frac{\partial^2}{\partial S'^2}V_f + (r_f - r_d)S'\frac{\partial}{\partial S'}V_f - r_f V_f$$
$$+ (r_f X' - r_d S') H((X' - S')^+ - V_f) = 0, \quad S' > 0, \; t \in [0,T),$$
$$V_f(S',T) = (X' - S')^+,$$

where in the equations (8.10) and (8.11) the American call and put options in currency d on currency f have strike price X, and where in the equations (8.12) and (8.13) the American call and put options in currency f on currency d have strike price X'.

We prove only the first of the symmetry relationships in (3.24) in the Black and Scholes market environment, since the proof of the second symmetry relationship in (3.24) in the Black and Scholes market environment is completely analogous.

Let V_d be the solution of the Cauchy problem for the semilinear evolution equation (8.8), that is, the value of the American option in currency d on currency f with inception time t, expiration time T, and with the payoff g_d.

We shall prove that V_f, such that $\mathbf{K} V_f = V_d$, is the solution of the Cauchy problem for the semilinear evolution equation (8.9) with $g_f = \mathbf{K} g_d$, that is, the value of the American option in currency f on currency d with inception time t, expiration time T, and with the payoff $g_f = \mathbf{K} g_d$.

Taking into account the fact that $\mathbf{K} = \mathbf{K}^{-1}$, it is easy to see, by substituting $\mathbf{K} V_f = V_d$ and $\mathbf{K} g_f = g_d$ into the Cauchy problem for the semilinear evolution equation (8.8), that V_f satisfies the Cauchy problem

$$\frac{\partial}{\partial t} \mathbf{K} V_f + L_d^{BS} \mathbf{K} V_f + (-(\frac{\partial}{\partial t} \mathbf{K} g_f + L_d^{BS} \mathbf{K} g_f))^+ H(\mathbf{K} g_f - \mathbf{K} V_f) = 0,$$
$$V_f|_{t=T} = g_f(\cdot, T),$$

where L_d^{BS} is the generator of the Black and Scholes market environment $\mathbf{V}_d^{BS}$, and where the arguments S and t of the functions in the semilinear evolution equation were suppressed for brevity.

Now, due to the symmetry relationships in (3.18) and due to the fact that the Kelvin transform $\mathbf{K}$ commutes with the operator of partial differentiation with respect to time t, the preceding Cauchy problem can be rewritten as

$$\mathbf{K} \frac{\partial}{\partial t} V_f + \mathbf{K} L_f^{BS} V_f + (-(\mathbf{K} \frac{\partial}{\partial t} g_f + \mathbf{K} L_f^{BS} g_f))^+ H(\mathbf{K} g_f - \mathbf{K} V_f) = 0,$$
$$V_f|_{t=T} = g_f(\cdot, T),$$

where L_f^{BS} is the generator of the Black and Scholes market environment $\mathbf{V}_f^{BS}$.

In turn, due to the definitions of the Kelvin transform $\mathbf{K}$, the Heaviside function H and the function $(\cdot)^+$, the preceding Cauchy problem can be rewritten as

$$\mathbf{K} \frac{\partial}{\partial t} V_f + \mathbf{K} L_f^{BS} V_f + \mathbf{K} ((-(\frac{\partial}{\partial t} g_f + L_f^{BS} g_f))^+ H(g_f - V_f)) = 0,$$
$$V_f|_{t=T} = g_f(\cdot, T).$$

Finally, acting by $\mathbf{K}^{-1}$ on both sides of the preceding equation we obtain the desired result that $V_f = \mathbf{K} V_d$ is the solution of the Cauchy problem for the semilinear evolution equation (8.9) with $g_f = \mathbf{K}g_d$, that is, the value of the American option in currency f on currency d with inception time t, expiration time T, and with the payoff $g_f = \mathbf{K}g_d$.

We note that the preceding proof with the help of the symmetry relationships (3.17) remains valid for the case of general market environments $\mathbf{V}_d$ and $\mathbf{V}_f$ generated by the families of linear operators

$$\mathbf{L}_d = \{\mathbf{L}_d(t) : \Pi \to \Pi \,|\, t \in \mathcal{T}\}$$

and

$$\mathbf{L}_f = \{\mathbf{L}_f(t) : \Pi \to \Pi \,|\, t \in \mathcal{T}\}.$$

We comment again that in this case the generators $\mathbf{L}_d^{BS}$ and $\mathbf{L}_f^{BS}$ of the Black and Scholes market environments $\mathbf{V}_d^{BS}$ and $\mathbf{V}_f^{BS}$ are replaced by the generators $\mathbf{L}_d(t)$ and $\mathbf{L}_f(t)$.

8.7 Approximation of American Options by Bermudan Options

As promised in Chapter 3, we will present a detailed analysis of the numerical significance of the fact that the symmetry relationships hold at all levels of accuracy in the numerical evaluation of foreign exchange American options in the Black and Scholes market environments $\mathbf{V}_d^{BS}$ and $\mathbf{V}_f^{BS}$ via approximations by the corresponding Bermudan options valued in the binomial market environments $\mathbf{V}_d^{B_n}$ and $\mathbf{V}_f^{B_n}$ restricted to the binomial trees Γ_d^n and Γ_f^n, that is, in the restricted binomial market environments $\hat{\mathbf{V}}_d^{B_n}$ and $\hat{\mathbf{V}}_f^{B_n}$. We will illustrate this fact using the example of such numerical evaluation of one of the most widely traded foreign exchange American options, namely American call and put options.

Denote the values of the Bermudan call and put options that approximate such American call and put options in currency d on currency f by $\hat{C}_d^n(t, T, S, X)$ and $\hat{P}_d^n(t, T, S, X)$. Similarly, denote the values of the Bermudan call and put options that approximate such American call and put options in currency f on currency d by $\hat{C}_f^n(t, T, S', X')$ and $\hat{P}_f^n(t, T, S', X')$. We comment that the exercise time sets of these Bermudan call and put options are equal to the trading time sets of the restricted binomial market environments $\hat{\mathbf{V}}_d^{B_n}$ and $\hat{\mathbf{V}}_f^{B_n}$.

Since, as we pointed out in Chapter 4, the binomial market environments $\mathbf{V}_d^{B_n}$ and $\mathbf{V}_f^{B_n}$ are exchange-rate homogeneous, their restrictions to the binomial trees Γ_d^n and Γ_f^n, that is, the restricted binomial market environments

$\hat{\mathbf{V}}_d^{B^n}$ and $\hat{\mathbf{V}}_f^{B^n}$, are also exchange-rate homogeneous in the sense that

$$\lambda \hat{C}_d^n(t, T, S, X) = \hat{C}_d^n(t, T, \lambda S, \lambda X),$$
$$\lambda \hat{P}_d^n(t, T, S, X) = \hat{P}_d^n(t, T, \lambda S, \lambda X),$$

$$\lambda \hat{C}_f^n(t, T, S', X') = \hat{C}_f^n(t, T, \lambda S', \lambda X'),$$
$$\lambda \hat{P}_f^n(t, T, S', X') = \hat{P}_f^n(t, T, \lambda S', \lambda X'),$$

for all nonnegative λ, where in these homogeneity relationships call and put options are respectively either European or Bermudan with arbitrary exercise time sets. Note that the call and put options in currency d on currency f on the left hand sides of these homogeneity relationships are valued in the binomial market environment $\mathbf{V}_d^{B^n}$ restricted to the binomial tree Γ_d^n, while the call and put options in currency d on currency f on the right hand sides of these homogeneity relationships are valued in the binomial market environment $\mathbf{V}_d^{B^n}$ restricted to the binomial tree

$$\lambda \Gamma_d^n = \Gamma_d^n \left((t_l, \lambda S_{lm}) : m = 1, \ldots, 2^l, l = 0, 1, \ldots, n \right).$$

Similarly, the call and put options in currency f on currency d on the left hand sides of these homogeneity relationships are valued in the binomial market environment $\mathbf{V}_f^{B^n}$ restricted to the binomial tree Γ_f^n, while the call and put options in currency f on currency d on the right hand sides of these homogeneity relationships are valued in the binomial market environment $\mathbf{V}_d^{B^n}$ restricted to the binomial tree

$$\lambda \Gamma_f^n = \Gamma_f^n \left((t_l, \lambda S_{lm}') : m = 1, \ldots, 2^l, l = 0, 1, \ldots, n \right).$$

Hence the symmetry relationships (3.22) and (3.23) for the Bermudan call and put options in the restricted binomial market environments $\hat{\mathbf{V}}_d^{B^n}$ and $\hat{\mathbf{V}}_f^{B^n}$ that approximate the American call and put options in the Black and Scholes market environments $\mathbf{V}_d^{BS}$ and $\mathbf{V}_f^{BS}$ can be rewritten as follows:

(8.14)
$$\hat{C}_d^n(t, T, S, X) = \hat{P}_f^n(t, T, X, S),$$
$$\hat{P}_d^n(t, T, S, X) = \hat{C}_f^n(t, T, X, S),$$

(8.15)
$$\hat{C}_f^n(t, T, S', X') = \hat{P}_d^n(t, T, X', S'),$$
$$\hat{P}_f^n(t, T, S', X') = \hat{C}_d^n(t, T, X', S'),$$

where, in (8.14), the Bermudan call and put options in currency d on currency f are valued in the binomial market environments $\mathbf{V}_d^{B_n}$ restricted to the binomial tree Γ_d^n, while the Bermudan call and put options in currency f on currency d are valued in the binomial market environments $\mathbf{V}_f^{B_n}$ restricted to the binomial tree

$$SX\,\Gamma_f^n = \Gamma_f^n\left((t_l, SX\,S_{lm}') : m = 1, \ldots, 2^l, l = 0, 1, \ldots, n\right)$$

and where, in (8.15), the Bermudan call and put options in currency f on currency d are valued in the binomial market environments $\mathbf{V}_f^{B_n}$ restricted to the binomial tree Γ_f^n, while the Bermudan call and put options in currency d on currency f are valued in the binomial market environments $\mathbf{V}_d^{B_n}$ restricted to the binomial tree

$$S'X'\,\Gamma_d^n = \Gamma_d^n\left((t_l, S'X'\,S_{lm}) : m = 1, \ldots, 2^l, l = 0, 1, \ldots, n\right).$$

The reason for the preceding rewriting of the symmetry relationships (3.22) and (3.23) in the form (8.14) and (8.15) is not just for sake of mathematical completeness. Rather, it is because the symmetry relationships (8.14) and (8.15) are more convenient for numerical evaluation of the American call and put options under consideration.

For sufficiently large n, typically of the order of 150 to 200, the values of the foreign exchange Bermudan call and put options in (8.14) and (8.15) serve as proxies for the values of the corresponding foreign exchange American call and put options in the Black and Scholes market environment. (See, for example, Jarrow and Rudd [11].) The justification for these values of Bermudan call and put options being such proxies is based on numerical experimentation that establishes numerical convergence of these values over the index n. The criterion for this numerical convergence is internal in the sense that it compares the values of these Bermudan call and put options for increasing values of the index n.

Since the criterion for this numerical convergence is internal, that is, based solely on the values of the Bermudan call and put options themselves, the fact that a desired level of accuracy is achieved on one side of the foreign exchange market implies, by the symmetry relationships (8.14) and (8.15), that the same level of accuracy is achieved on the opposite side of the market. Therefore, once we have valued an American call or put option in the Black and Scholes market environment on one side of the foreign exchange market with any desired level of accuracy, we can instantly value its counterpart on the opposite side of the market by the symmetry relationships (8.14) and (8.15) with the same level of accuracy.

This is illustrated in Figure 8.2. The shades of gray above the main diagonal represent the number of steps n necessary to achieve two decimal places of accuracy in the approximate evaluation of the American call options $C_d^{BS}(t, T, S, X)$ in currency d on currency f in the Black and Scholes market environment as a Bermudan call option $\hat{C}_d^n(t, T, S, X)$ in (8.14). The horizontal axis represents time to expiration $T-t$ from zero to six months and the vertical axis represents the exchange rate S in units of the strike X. Similarly, the shades of gray below the main diagonal represent the number of steps n necessary to achieve two decimal places of accuracy in the approximate evaluation of the American put options $P_f^{BS}(t, T, S', X')$ in currency f on currency d in the Black and Scholes market environment as a Bermudan put option $\hat{P}_d^n(t, T, S', X')$ in (8.14) with $S' = X$ and $X' = S$. The vertical axis represents time to expiration $T-t$ from zero to six months and the horizontal axis represents the exchange rate S' in units of the strike X'. The symmetry in the figure about the main diagonal is due to the first symmetry relationship in (8.14) for Bermudan call and put options in the restricted binomial market environment and reflects the fact that once one has valued an American call or put option in the Black and Scholes market environment via these Bermudan call or put options on one side of the foreign exchange market with any desired level of accuracy, one can instantly value its counterpart on the opposite side of the market with the same level of accuracy.

8.8   Symmetry at all Levels of Approximation

In general, in order to compute the value of a foreign exchange option that does not have a closed-form solution, which includes almost every option, some approximation scheme has to be employed. In this case, the validity of the symmetry relationships at all levels of approximation becomes especially important when one does not have any a priori estimates for the precision of the chosen approximation scheme. One is forced to judge the precision of the chosen approximation scheme based on internal criteria. Therefore, the fact that a desired level of accuracy is achieved on one side of the foreign exchange market implies, by the symmetry relationships in the foreign exchange market, that the same level of accuracy is achieved on the opposite side of the market. This means that once this foreign exchange option is valued on one side of the market, then automatically its counterpart is valued on the other side of the market with the same level of accuracy.

This completes the proofs of the symmetry relationships for foreign exchange Bermudan and American options with general time-dependent payoffs

Figure 8.2: **Approximate evaluation of foreign exchange American call and put options using the symmetry relationships in the binomial market environment.** The symmetry in the figure about the main diagonal is due to the first symmetry relationship in (8.14) for Bermudan call and put options in the binomial market environment, or more precisely, in the restricted binomial market environment and reflects the fact that once one has valued an American call or put option in the Black and Scholes market environment via these Bermudan call or put options on one side of the foreign exchange market with any desired level of accuracy, one can instantly value its counterpart on the opposite side of the market with the same level of accuracy.)

in a general market environment satisfying the intervention condition. In the next chapter, we use these symmetry relationships to prove the symmetry relationships for foreign exchange barrier options presented in Chapter 4.

Chapter 9

Validity of the Symmetry Relationships for Barrier Options

In this Chapter we prove the symmetry relationships presented in Chapter 4 for one of the major classes of foreign exchange exotic options, namely barrier options in general market environments V_d and V_f satisfying the intervention condition. We consider barrier options in the generality of an arbitrary time-dependent barrier which can be activated not only on the entire life of the option, but on any discrete set of times during its life. Furthermore, we consider barrier options of European, Bermudan and American styles with the underlying European, Bermudan and American options having general and, in the latter two cases, time-dependent payoffs. Our proof is based on the fact shown in [18], that such barrier options can be viewed as Bermudan options with certain payoffs for discretely activated barriers or as American options with certain payoffs for continuously activated barriers. This fact enables us to use the symmetry relationships presented in Chapter 3 and proved in the previous Chapter to derive the symmetry relationships for such barrier options.

9.1 Barrier Options: General Case

In this section we prove the symmetry relationships for barrier options in the foreign exchange market presented in Chapter 4. We start with the symme-

115

try relationships (4.13)–(4.16), proving only their first relationships, since the proofs of the second relationships are completely analogous. For the same reason, we give proofs only in the case when the barrier activation set E_b is discrete. In order to do this, we need further preparation.

Firstly, we state the following obvious identity:

$$(9.1) \qquad (\mathbf{K}(g\,h))(S) = g(1/S)(\mathbf{K}h)(S), \quad S > 0,$$

where g and h are in Π and $g\,h$ stands for the product of functions g and h.

Secondly, we note that the uniform notation in (4.2) and (4.4) permits the symmetry relationships (3.5), (3.19) and (3.24) for European, Bermudan, and American options to be rewritten as

$$(9.2) \qquad \begin{aligned} &\mathcal{O}_d(t,T,E_\mathcal{O},g_d) = \mathbf{K}\,\mathcal{O}_f(t,T,E_\mathcal{O},\mathbf{K}g_d), \quad g_d : E_\mathcal{O} \to \Pi_+, \\ &\mathcal{O}_f(t,T,E_\mathcal{O},g_f) = \mathbf{K}\,\mathcal{O}_d(t,T,E_\mathcal{O},\mathbf{K}g_f), \quad g_f : E_\mathcal{O} \to \Pi_+. \end{aligned}$$

Now, for the payoff g_d^D in (4.6) of the Bermudan option with value $\mathcal{B}_d(t,T,E_b,g_d^D)$, which is equivalent in the sense of (4.5) to the down and in barrier option with value $DI_d(t,T,E_b,b_d,E_\mathcal{O},g_d)$, we have the following chain of equalities:

$$(9.3) \qquad \begin{aligned} (\mathbf{K}\,g_{d,\tau}^D)(S_\tau) &= (\mathbf{K}\,H(b_d(\tau) - \cdot)\,\mathcal{O}_d(\tau,T,E_\mathcal{O},g_d)(\cdot))\,(S_\tau) \\ &= H(b_d(\tau) - 1/S_\tau)\,(\mathbf{K}\,\mathcal{O}_d(\tau,T,E_\mathcal{O},g_d))\,(S_\tau) \\ &= H(S_\tau - 1/b_d(\tau))\mathcal{O}_f(\tau,T,E_\mathcal{O},\mathbf{K}g_d)(S_\tau), \end{aligned}$$

where the first equality is due to the definition of g_d^D in (4.6), the second equality is due to the identity (9.1), and the last equality is due the definition of the Heaviside function H and the first symmetry relationship in (9.2) along with the fact that $\mathbf{K}^2$ is the identity operator on Π.

Thus the proof of the first symmetry relationship in (4.13) can be completed as follows:

$$\begin{aligned} DI_d&(t,T,E_b,b_d,E_\mathcal{O},g_d) \\ &= \mathcal{B}_d\,(t,T,E_b,H(b_d(\circ) - \cdot)\,\mathcal{O}_d(\circ,T,E_\mathcal{O},g_d)(\cdot)) \\ &= \mathbf{K}\,\mathcal{B}_f\,(t,T,E_b,\mathbf{K}H(b_d(\circ) - \cdot)\,\mathcal{O}_d(\circ,T,E_\mathcal{O},g_d)(\cdot)) \\ &= \mathbf{K}\,\mathcal{B}_f\,(t,T,E_b,H(\cdot - 1/b_d(\circ))\,(\mathcal{O}_f(\circ,T,E_\mathcal{O},\mathbf{K}g_d)(\cdot)) \\ &= \mathbf{K}\,UI_f(t,T,E_b,1/b_d,E_\mathcal{O},\mathbf{K}g_d), \end{aligned}$$

where the first equality is due to the relationship (4.5), the second equality is due to the first symmetry relationship in (3.19), the third equality is due to

the preceding relationship (9.3), and the fourth equality is due to the relationship (4.11). We comment that here and below in this section, the symbol $\circ$
stands for the current time variable and the symbol $\cdot$ stands, as usual, for the
exchange rate variable at this current time.

Similarly, we prove the first symmetry relationship in (4.14). For the pay-
off g_d^U in (4.10) of the Bermudan option with value $\mathcal{B}_d(t, T, E_b, g_d^U)$, which is
equivalent in the sense of (4.9) to the up and in barrier option with value
$UI_d(t, T, E_b, b_d, E_{\mathcal{O}}, g_d)$, we have the following chain of equalities:

$$(\mathbf{K}\, g_{d,\tau}^U)(S_\tau) = (\mathbf{K}\, H(\cdot - b_d(\tau))\, \mathcal{O}_d(\tau, T, E_{\mathcal{O}}, g_d)(\cdot))\, (S_\tau)$$
$$= H(1/S_\tau - b_d(\tau))\, (\mathbf{K}\, \mathcal{O}_d(\tau, T, E_{\mathcal{O}}, g_d))\, (S_\tau)$$
$$= H(1/b_d(\tau) - S_\tau)\mathcal{O}_f(\tau, T, E_{\mathcal{O}}, \mathbf{K}g_d)(S_\tau).$$

Thus the proof of the first symmetry relationship in (4.14) can be completed
as follows:

$$UI_d(t, T, E_b, b_d, E_{\mathcal{O}}, g_d)$$
$$= \mathcal{B}_d\left(t, T, E_b, H(\cdot - b_d(\circ))\, \mathcal{O}_d(\circ, T, E_{\mathcal{O}}, g_d)(\cdot)\right)$$
$$= \mathbf{K}\, \mathcal{B}_f\left(t, T, E_b, \mathbf{K}H(\cdot - b_d(\circ))\, \mathcal{O}_d(\circ, T, E_{\mathcal{O}}, g_d)(\cdot)\right)$$
$$= \mathbf{K}\, \mathcal{B}_f\left(t, T, E_b, H(1/b_d(\circ) - \cdot)\, (\mathcal{O}_f(\circ, T, E_{\mathcal{O}}, \mathbf{K}g_d)(\cdot)\right)$$
$$= \mathbf{K}\, DI_f(t, T, E_b, 1/b_d, E_{\mathcal{O}}, \mathbf{K}g_d).$$

Next we prove the first symmetry relationship in (4.15) via the following
chain of equalities:

$$DO_d(t, T, E_b, b_d, E_{\mathcal{O}}, g_d)$$
$$= \mathcal{O}_d(t, T, E_{\mathcal{O}}, g_d) - DI_d(t, T, E_b, b_d, E_{\mathcal{O}}, g_d)$$
$$= \mathbf{K}\mathcal{O}_f(t, T, E_{\mathcal{O}}, \mathbf{K}g_d) - \mathbf{K}\, UI_f(t, T, E_b, 1/b_d, E_{\mathcal{O}}, \mathbf{K}g_d)$$
$$= \mathbf{K}UO_f(t, T, E_b, 1/b_d, E_{\mathcal{O}}, \mathbf{K}g_d),$$

where the first equality is just the first relationship in (4.1), the second equality
is due to the first symmetry relationship in (9.2) and to the first symmetry
relationship in (4.13), and the last equality is due to the second relationship
in (4.3).

Similarly, we prove the first symmetry relationship in (4.16) via the following chain of equalities:

$$UO_d(t, T, E_b, b_d, E_{\mathcal{O}}, g_d)$$
$$= \mathcal{O}_d(t, T, E_{\mathcal{O}}, g_d) - UI_d(t, T, E_b, b_d, E_{\mathcal{O}}, g_d)$$
$$= \mathbf{K}\mathcal{O}_f(t, T, E_{\mathcal{O}}, \mathbf{K}g_d) - \mathbf{K}\, DI_f(t, T, E_b, 1/b_d, E_{\mathcal{O}}, \mathbf{K}g_d)$$
$$= \mathbf{K}DO_f(t, T, E_b, 1/b_d, E_{\mathcal{O}}, \mathbf{K}g_d).$$

This completes the proofs of the symmetry relationships in (4.13)–(4.16) for barrier options in the general case.

9.2 Barrier Call and Put Options

Our next goal is to prove the particular cases (4.17)–(4.24) for barrier call and put options of the symmetry relationships in (4.13)–(4.16) for general barrier options.

We prove only the first symmetry relationships in (4.17), (4.19), (4.21) and (4.23) since the proofs of the second relationships are completely analogous. For the same reason, we give proofs only in the case when the barrier activation set E_b is discrete. In order to do this, we need further preparation.

Firstly, the uniform notation in (4.2) and (4.4) permits the homogeneity relationships (7.1), (8.3) and (8.7) for European, Bermudan and American options to be rewritten as

$$(9.4) \qquad \begin{aligned} \mathcal{O}_d(t, T, E_{\mathcal{O}}, \lambda g_d) &= \lambda\, \mathcal{O}_d(t, T, E_{\mathcal{O}}, g_d), \\ \mathcal{O}_f(t, T, E_{\mathcal{O}}, \lambda g_f) &= \lambda\, \mathcal{O}_f(t, T, E_{\mathcal{O}}, g_f), \end{aligned}$$

for all nonnegative λ.

Secondly, we need the analogous homogeneity relationships for barrier options of European, Bermudan and American style, namely,

$$(9.5) \qquad \begin{aligned} DI_d(t, T, E_b, b_d, E_{\mathcal{O}}, \lambda g_d) &= \lambda\, DI_d(t, T, E_b, b_d, E_{\mathcal{O}}, g_d), \\ DI_f(t, T, E_b, b_f, E_{\mathcal{O}}, \lambda g_f) &= \lambda\, DI_f(t, T, E_b, b_f, E_{\mathcal{O}}, g_f), \end{aligned}$$

$$(9.6) \qquad \begin{aligned} UI_d(t, T, E_b, b_d, E_{\mathcal{O}}, \lambda g_d) &= \lambda\, UI_d(t, T, E_b, b_d, E_{\mathcal{O}}, g_d), \\ UI_f(t, T, E_b, b_f, E_{\mathcal{O}}, \lambda g_f) &= \lambda\, UI_f(t, T, E_b, b_f, E_{\mathcal{O}}, g_f), \end{aligned}$$

$$(9.7) \qquad \begin{aligned} DO_d(t, T, E_b, b_d, E_{\mathcal{O}}, \lambda g_d) &= \lambda\, DO_d(t, T, E_b, b_d, E_{\mathcal{O}}, g_d), \\ DO_f(t, T, E_b, b_f, E_{\mathcal{O}}, \lambda g_f) &= \lambda\, DO_f(t, T, E_b, b_f, E_{\mathcal{O}}, g_f), \end{aligned}$$

$$(9.8) \qquad \begin{aligned} UO_d(t, T, E_b, b_d, E_{\mathcal{O}}, \lambda g_d) &= \lambda\, UO_d(t, T, E_b, b_d, E_{\mathcal{O}}, g_d), \\ UO_f(t, T, E_b, b_f, E_{\mathcal{O}}, \lambda g_f) &= \lambda\, UO_f(t, T, E_b, b_f, E_{\mathcal{O}}, g_f), \end{aligned}$$

for all nonnegative λ.

These homogeneity relationships were proved in [18]. Here, we will prove only the first homogeneity relationships in (9.5)–(9.8) since the proofs of the second relationships are analogous. For the same reason, we give proofs only in the case when the barrier activation set E_b is discrete.

The proof of the first relationship in (9.5) follows from the chain of equalities:

$$
\begin{aligned}
DI_d(t,T,E_b,b_d,E_{\mathcal{O}},\lambda g_d) &= \mathcal{B}_d(t,T,E_b,H(b_d(\circ)-\cdot)\mathcal{O}_d(\circ,T,E_{\mathcal{O}},\lambda g_d)(\cdot)) \\
&= \mathcal{B}_d(t,T,E_b,H(b_d(\circ)-\cdot)\lambda\mathcal{O}_d(\circ,T,E_{\mathcal{O}},g_d)(\cdot)) \\
&= \lambda\mathcal{B}_d(t,T,E_b,H(b_d(\circ)-\cdot)\mathcal{O}_d(\circ,T,E_{\mathcal{O}},g_d)(\cdot)) \\
&= \lambda DI_d(t,T,E_b,b_d,E_{\mathcal{O}},g_d),
\end{aligned}
$$

where the first and the last equalities are due to the relationship (4.5), the second equality is due to the first relationship in (9.4), and the third equality is due to the first relationship in (8.3).

Similarly, the proof of the first relationship in (9.6) follows from the chain of equalities:

$$
\begin{aligned}
UI_d(t,T,E_b,b_d,E_{\mathcal{O}},\lambda g_d) &= \mathcal{B}_d(t,T,E_b,H(\cdot-b_d(\circ))\mathcal{O}_d(\circ,T,E_{\mathcal{O}},\lambda g_d)(\cdot)) \\
&= \mathcal{B}_d(t,T,E_b,H(\cdot-b_d(\circ))\lambda\mathcal{O}_d(\circ,T,E_{\mathcal{O}},g_d)(\cdot)) \\
&= \lambda\mathcal{B}_d(t,T,E_b,H(\cdot-b_d(\circ))\mathcal{O}_d(\circ,T,E_{\mathcal{O}},g_d)(\cdot)) \\
&= \lambda UI_d(t,T,E_b,b_d,E_{\mathcal{O}},g_d).
\end{aligned}
$$

Now we turn to the proof of the first relationship in (9.7), which can be completed by the chain of equalities:

$$
\begin{aligned}
DO_d(t,T,E_b,b_d,E_{\mathcal{O}},\lambda g_d) &= \mathcal{O}_d(t,T,E_{\mathcal{O}},\lambda g_d) - DI_d(t,T,E_b,b_d,E_{\mathcal{O}},\lambda g_d) \\
&= \lambda\mathcal{O}_d(t,T,E_{\mathcal{O}},g_d) - \lambda DI_d(t,T,E_b,b_d,E_{\mathcal{O}},g_d) \\
&= \lambda DO_d(t,T,E_b,b_d,E_{\mathcal{O}},g_d),
\end{aligned}
$$

where the first and last equalities are due to the first relationship in (4.1), and the second equality is due to the first relationship in (9.4) and the first relationship in (9.5).

Similarly, the proof of the first relationship in (9.8) can be completed by the chain of equalities:

$$
\begin{aligned}
UO_d(t,T,E_b,b_d,E_{\mathcal{O}},\lambda g_d) &= \mathcal{O}_d(t,T,E_{\mathcal{O}},\lambda g_d) - UI_d(t,T,E_b,b_d,E_{\mathcal{O}},\lambda g_d) \\
&= \lambda\mathcal{O}_d(t,T,E_{\mathcal{O}},g_d) - \lambda UI_d(t,T,E_b,b_d,E_{\mathcal{O}},g_d) \\
&= \lambda UO_d(t,T,E_b,b_d,E_{\mathcal{O}},g_d).
\end{aligned}
$$

We now turn to the proof of the first symmetry relationship in (4.17):

$$DIC_d(t, T, E_b, b_d, E_\mathcal{O}, S, X) = DI_d(t, T, E_b, b_d, E_\mathcal{O}, (\cdot - X)^+)(S)$$
$$= \mathbf{K}\, UI_f(t, T, E_b, 1/b_d, E_\mathcal{O}, \mathbf{K}\, (\cdot - X)^+)(S)$$
$$= \mathbf{K}\, UI_f(t, T, E_b, 1/b_d, E_\mathcal{O}, X(1/X - \cdot)^+)(S)$$
$$= X\, \mathbf{K}\, UI_f(t, T, E_b, 1/b_d, E_\mathcal{O}, (1/X - \cdot)^+)(S)$$
$$= XS\, UIP_f(t, T, E_b, 1/b_d, E_\mathcal{O}, 1/S, 1/X),$$

where the first equality is just the definition of

$$DIC_d(t, T, E_b, b, E_\mathcal{O}, S, X)$$

in (4.25), the second equality is due to the first symmetry relationship in (4.13), the third equality is due the first symmetry relationship in (3.2), the fourth equality is due to the second homogeneity relationship in (9.6) and the linearity of $\mathbf{K}$, and the last equality is due to the definition of

$$UIP_f(t, T, E_b, b, E_\mathcal{O}, S', X')$$

in (4.26) and the definition of the Kelvin transform.

Similarly, the proof of the first symmetry relationship in (4.19) is completed by the following chain of equalities:

$$UIC_d(t, T, E_b, b_d, E_\mathcal{O}, S, X) = UI_d(t, T, E_b, b_d, E_\mathcal{O}, (\cdot - X)^+)(S)$$
$$= \mathbf{K}\, DI_f(t, T, E_b, 1/b_d, E_\mathcal{O}, \mathbf{K}\, (\cdot - X)^+)(S)$$
$$= \mathbf{K}\, DI_f(t, T, E_b, 1/b_d, E_\mathcal{O}, X(1/X - \cdot)^+)(S)$$
$$= X\mathbf{K}\, DI_f(t, T, E_b, 1/b_d, E_\mathcal{O}, (1/X - \cdot)^+)(S)$$
$$= XS\, DIP_f(t, T, E_b, b_d, E_\mathcal{O}, 1/S, 1/X).$$

Next we prove the first relationship in (4.21):

$$DOC_d(t, T, E_b, b_d, E_\mathcal{O}, S, X) = DO_d(t, T, E_b, b_d, E_\mathcal{O}, (\cdot - X)^+)(S)$$
$$= \mathbf{K}\, UO_f(t, T, E_b, 1/b_d, E_\mathcal{O}, \mathbf{K}\, (\cdot - X)^+)(S)$$
$$= \mathbf{K}\, UO_f(t, T, E_b, 1/b_d, E_\mathcal{O}, X(1/X - \cdot)^+)(S)$$
$$= X\, \mathbf{K}\, UO_f(t, T, E_b, 1/b_d, E_\mathcal{O}, (1/X - \cdot)^+)(S)$$
$$= XS\, UOP_f(t, T, E_b, 1/b_d, E_\mathcal{O}, 1/S, 1/X),$$

where the first equality is just the definition of

$$DOC_d(t, T, E_b, b, E_\mathcal{O}, S, X)$$

in (4.27), the second equality is due to the first symmetry relationship in (4.15), the third equality is due to the first symmetry relationship in (3.2), the fourth equality is due to the second homogeneity relationship in (9.8) and the linearity of $\mathbf{K}$, and the last equality is due to the definition of

$$UOP_f(t, T, E_b, b, E_{\mathcal{O}}, S', X')$$

in (4.28) and to the definition of the Kelvin transform.

Similarly, the proof of the first symmetry relationship in (4.23) is completed by the chain of equalities:

$$
\begin{aligned}
UOC_d(t, T, E_b, b_d, E_{\mathcal{O}}, S, X) &= UO_d(t, T, E_b, b_d, E_{\mathcal{O}}, (\cdot - X)^+)(S) \\
&= \mathbf{K}\, DO_f(t, T, E_b, 1/b_d, E_{\mathcal{O}}, \mathbf{K}\,(\cdot - X)^+)(S) \\
&= \mathbf{K}\, DO_f(t, T, E_b, 1/b_d, E_{\mathcal{O}}, X(1/X - \cdot)^+)(S) \\
&= X\mathbf{K}\, DO_f(t, T, E_b, 1/b_d, E_{\mathcal{O}}, (1/X - \cdot)^+)(S) \\
&= XS\, DOP_f(t, T, E_b, 1/b_d, E_{\mathcal{O}}, 1/S, 1/X).
\end{aligned}
$$

This completes the proofs of the symmetry relationships for foreign exchange barrier options.

Chapter 10

Validity of the Symmetry Relationships for Options with Consistently Smoothed Payoffs

In this chapter we complete our presentation of formal mathematical proofs of the symmetry relationships presented in Chapters 3 through 5 by proving the symmetry relationships (5.5) and (5.6) for the consistently smoothed call and put payoffs and the symmetry relationships (5.7) and (5.8) for European, Bermudan and American options with these payoffs presented in Chapter 5.

10.1   The Consistently Smoothed Call and Put Payoffs

We now turn to the proofs of the symmetry relationships (5.5) and (5.6) for the consistently smoothed call and put payoffs

$$g_d^c(L_d, U_d, X), \ g_d^p(L_d, U_d, X), \ g_f^c(L_f, U_f, X'), \ \text{and} \ g_f^c(L_f, U_f, X')$$

introduced in Section 5.4.

We prove only the first symmetry relationship in (5.5), since the proof of the second symmetry relationship in (5.5) and the proofs of the symmetry relationships in (5.6) are analogous.

We consider two cases:

Case 1. $0 < 1/S' \leq L_d$ and $1/S' \geq U_d$.

The following chain of equalities establishes the proof in this case:

$$
\begin{aligned}
(\mathbf{K} g_d^c(L_d, U_d, X))(S') &= S' \, g_d^c(L_d, U_d, X))(1/S') \\
&= S' \, (1/S' - X)^+ \\
&= X \, (1/X - S')^+ \\
&= X \, g_f^p(1/U_d, 1/L_d, 1/X)(S'),
\end{aligned}
$$

where the first equality is due to the definition of the Kelvin transform, the second and the last equalities are due to the definitions of $g_d^c(L_d, U_d, X)$ and $g_f^p(L_f, U_f, X')$ and the third equality is due to the first symmetry relationship in (3.2).

Case 2. $L_d \leq 1/S' \leq U_d$.

The following chain of equalities establishes the proof in this case:

$$
\begin{aligned}
(\mathbf{K} \, g_d^c(L_d, U_d, X))(S') &= S' \, g_d^c(L_d, U_d, X))(1/S') \\
&= \frac{S' \beta_d (1/S' - L_d)^{2\alpha_d + 1}}{(1/S')^{\alpha_d}} \\
&= \frac{\beta_d L_d^{2\alpha_d + 1} (1/L_d - S')^{2\alpha_d + 1}}{S'^{\alpha_d}} \\
&= \frac{(L_d U_d)^{1/2} \delta_f (1/L_d - S')^{2\gamma_f + 1}}{S'^{\gamma_f}} \\
&= X \, g_f^p(1/U_d, 1/L_d, 1/X)(S'),
\end{aligned}
$$

where the first equality is due to the definition of the Kelvin transform, the second and the last equalities are due to the definitions of $g_d^c(L_d, U_d, X)$ and $g_f^p(L_f, U_f, X')$, the third equality follows from elementary algebra, and the fourth equality follows from the obvious identities:

$$
\begin{aligned}
\alpha_d(L_d, U_d) &= \gamma_f(1/U_d, 1/L_d), \\
\beta_d(L_d, U_d) &= \sqrt{L_d U_d} L_d^{-(2\alpha_d + 1)} \delta_f(1/U_d, 1/L_d).
\end{aligned}
$$

10.2   Options with the Consistently Smoothed Call and Put Payoffs

In this section we prove the symmetry relationships (5.7) and (5.8) for European, Bermudan and American options with the consistently smoothed call and put payoffs

$$g_d^c(L_d, U_d, X), \; g_d^p(L_d, U_d, X), \; g_f^c(L_f, U_f, X'), \; \text{and } g_f^c(L_f, U_f, X').$$

We prove only the first symmetry relationship in (5.7) since the proof of the second symmetry relationship in (5.7) and the proofs of the symmetry relationships in (5.8) are analogous.

The proof follows from the chain of equalities:

$$
\begin{aligned}
C_d(t, T, S, X, L_d, U_d) &= \mathcal{O}_d(t, T, E_\mathcal{O}, g_d^c(L_d, U_d, X))(S) \\
&= (\mathbf{K}\, \mathcal{O}_f(t, T, E_\mathcal{O}, \mathbf{K}\, g_d^c(L_d, U_d, X)))(S) \\
&= (\mathbf{K}\, \mathcal{O}_f(t, T, E_\mathcal{O}, X\, g_f^p(1/U_d, 1/L_d, 1/X)))(S) \\
&= X\, (\mathbf{K}\, \mathcal{O}_f(t, T, E_\mathcal{O}, g_f^p(1/U_d, 1/L_d, 1/X)))(S) \\
&= SX\, P_f(t, T, 1/S, 1/X, 1/U_d, 1/L_d),
\end{aligned}
$$

where the first equality is due to the definition of $C_d(t, T, S, X, L_d, U_d)$, the second equality is due to the first symmetry relationship in (9.2), the third equality is due to the first symmetry relationship in (5.5), the fourth equality is due to the first homogeneity relationship in (9.4) and the linearity of the Kelvin transform, and the last equality is due to the definitions of $P_f(t, T, S', X', L_f, U_f)$ and the Kelvin transform.

We stress here again the crucial role played by the fact mentioned in Chapter 5, that the symmetry relationships (5.7) and (5.8) are based on the symmetry relationships (9.2) for European, Bermudan and American options with general payoffs rather than just for such call and put options. This means that in order to use symmetry relationships to guide the choice of new payoffs that smooth the corners for the payoffs of European, Bermudan, and American options, even for the case of such call and put options, one is forced to consider the symmetry relationships for European, Bermudan, and American options with payoffs of a general type.

This completes the proofs of the symmetry relationships for the consistently smoothed call and put payoffs and for European, Bermudan, and American options with these payoffs. In turn, this completes the formal mathematical proofs of the symmetry relationships presented in Chapters 3 through 5.

Bibliography

[1] F. Black and M. Scholes. The pricing of options and corporate liabilities. *Journal of Political Economy*, 81:637–657, May-Jun 1973.

[2] J. C. Cox, S. A. Ross, and M. Rubinstein. Option pricing: a simplified approach. *Journal of Financial Economics*, 7:229–263, Sep 1979.

[3] C. Davidson. Wider horizons. *Risk*, pages 55–59, April 1996.

[4] A. Friedman. *Partial Differential Equations of Parabolic Type*. Prentice-Hall, Inc., Englewood Cliffs, New Jersey, 1964.

[5] V. Frishling. Barrier rife. *Australia and New Zealand: A Supplement to Risk Magazine*, pages 23–24, August 1997.

[6] M. B. Garman. Towards a semigroup pricing theory. *Journal of Finance*, 40:847–861, July 1985.

[7] M. B. Garman and S. W. Kohlhagen. Foreign currency option values. *Journal of International Money and Finance*, 2:231–237, Dec 1983.

[8] R. Gibson. *Option Valuation*. McGraw-Hill, New York, 1991.

[9] J. O. Grabbe. The pricing of call and put options on foreign exchange. *Journal of International Money and Finance*, 2:239–253, Dec 1983.

[10] I. Hart and M. Ross. Striking continuity. *Risk*, 7(6):53–56, Jun 1994.

[11] R. Jarrow and A. Rudd. *Option pricing*. Richard D. Irwin, Homewood, Illinois, 1983.

[12] V. A. Kholodnyi. Analysis of the semilinear evolution equation for american options based on the theory of semigroups with multivalued generators. *Preprint, Integrated Energy Services, Inc.*, 1995.

[13] V. A. Kholodnyi. Beliefs-preferences gauge symmetry group and replication of contingent claims in a general market environment. *Preprint, Integrated Energy Services, Inc.*, 1995.

[14] V. A. Kholodnyi. Convergence of bermudan options to american options based on the theory of semigroups with multivalued generators. *Preprint, Integrated Energy Services, Inc.*, 1995.

[15] V. A. Kholodnyi. Convergence of discretely activated universal contingent claims to continuously activated universal contingent claims based on monotonicity. *Preprint, Integrated Energy Services, Inc.*, 1995.

[16] V. A. Kholodnyi. Convergence of discretely activated universal contingent claims to continuously activated universal contingent claims based on the theory of semigroups. *Preprint, Integrated Energy Services, Inc.*, 1995.

[17] V. A. Kholodnyi. Nonlinear partial differential equation for American options. *Preprint, Integrated Energy Services, Inc.*, 1995.

[18] V. A. Kholodnyi. On the linearity of Bermudan and American options with general time-dependent payoffs in partial semimodules. *Preprint, Integrated Energy Services, Inc.*, 1995.

[19] V. A. Kholodnyi. On weighted function spaces related to the Cauchy problem for the heat equation arising in valuation of contingent claims in the Black and Scholes market environment. *Preprint, Integrated Energy Services, Inc.*, 1995.

[20] V. A. Kholodnyi. Semilinear evolution equation for general derivative contracts. *Preprint, Integrated Energy Services, Inc.*, 1995.

[21] V. A. Kholodnyi. Semilinear evolution equation for universal contingent claims. *Preprint, Integrated Energy Services, Inc.*, 1995.

[22] V. A. Kholodnyi. Universal contingent claims. *Preprint, Integrated Energy Services, Inc.*, 1995.

[23] V. A. Kholodnyi. Approximation of foreign exchange options preserving the foreign exchange option symmetry. *Preprint, Integrated Energy Services, Inc.*, 1996.

[24] V. A. Kholodnyi. Foreign exchange option symmetry in the exchange-rate homogeneous market environment. *Preprint, Integrated Energy Services, Inc.*, 1996.

[25] V. A. Kholodnyi. Foreign exchange symmetry for universal contingent claims. *Preprint, Integrated Energy Services, Inc.*, 1996.

[26] V. A. Kholodnyi and J. F. Price. Foreign exchange option symmetry in a general market environment. *Preprint, Integrated Energy Services, Inc.*, 1996.

[27] V. A. Kholodnyi and J. F. Price. Foreign exchange option symmetry in a multiple currency general market environment. *Preprint, Integrated Energy Services, Inc.*, 1996.

[28] V. A. Kholodnyi and J. F. Price. Foreign exchange option symmetry based on domestic-foreign payoff invariance. In J. F. Marshall and R. J. Marks, editors, *Conference on Computational Intelligence for Financial Engineering*, pages 164–170. IEEE, 1997.

[29] V. A. Kholodnyi and J. F. Price. Graph theoretic formalism for foreign exchange option markets. In C. A. Gorini, editor, *Geometry at Work: A Collection of Papers in Applied Geometry*. Mathematics Association of America, 1997. To appear.

[30] R. C. Merton. Theory of rational option pricing. *Bell Journal of Economics and Management Science*, 4:141–183, 1973.

[31] J. F. Price. Horticulture for binomial trees. *Preprint, Maharishi University of Management*, 1995.

[32] J. F. Price. Optional mathematics is not optional. *Notices of the American Mathematical Society*, 43:964–971, Sep 1996.

[33] P. Ritchken. *Options: Theory, Strategy, and Applications*. Harper Collins, 1987.

[34] V. S. Vladimirov. *Equations of Mathematical Physics*. Nauka, Moscow, 1988.

[35] P. Wilmott, J. Dewynne, and S. Howison. *Option Pricing: Mathematical Models and Computation*. Oxford Financial Press, Oxford, UK, 1993.

Index